Cambridge Elements ⎯

Elements in Emerging Theories and Technologies
in Metamaterials
edited by
Tie Jun Cui
Southeast University, China
John B. P
Imperial Coll

EPSILON-NEAR-ZERO METAMATERIALS

Yue Li
Tsinghua University, China
Ziheng Zhou
Tsinghua University, China
Yijing He
Tsinghua University, China
Hao Li
Tsinghua University, China

CAMBRIDGE
UNIVERSITY PRESS

CAMBRIDGE
UNIVERSITY PRESS

University Printing House, Cambridge CB2 8BS, United Kingdom

One Liberty Plaza, 20th Floor, New York, NY 10006, USA

477 Williamstown Road, Port Melbourne, VIC 3207, Australia

314–321, 3rd Floor, Plot 3, Splendor Forum, Jasola District Centre,
New Delhi – 110025, India

103 Penang Road, #05–06/07, Visioncrest Commercial, Singapore 238467

Cambridge University Press is part of the University of Cambridge.

It furthers the University's mission by disseminating knowledge in the pursuit of
education, learning, and research at the highest international levels of excellence.

www.cambridge.org
Information on this title: www.cambridge.org/9781009124416
DOI: 10.1017/9781009128339

First published 2021

A catalogue record for this publication is available from the British Library.

ISBN 978-1-009-12441-6 Paperback
ISSN 2399-7486 (online)
ISSN 2514-3875 (print)

Cambridge University Press has no responsibility for the persistence or accuracy
of URLs for external or third-party internet websites referred to in this publication
and does not guarantee that any content on such websites is, or will remain,
accurate or appropriate.

Epsilon-Near-Zero Metamaterials

Elements in Emerging Theories and Technologies in Metamaterials

DOI: 10.1017/9781009128339
First published online: December 2021

Yue Li
Tsinghua University, China

Ziheng Zhou
Tsinghua University, China

Yijing He
Tsinghua University, China

Hao Li
Tsinghua University, China

Author for correspondence: Yue Li, lyee@tsinghua.edu.cn

Abstract: This Element introduces the unusual wave phenomena arising from an extremely small optical refractive index and sheds light on their underlying mechanisms, with a primary focus on the basic concepts and fundamental wave physics. The authors reveal the exciting applications of epsilon-near-zero (ENZ) metamaterials, which have profound impacts over a wide range of fields of science and technology. The sections are organized as follows: in Section 2, the authors demonstrate the extraordinary wave properties in ENZ metamaterials, analyzing the unique wave dynamics and the resulting effects. Section 3 is dedicated to introducing various realization methods of the ENZ metamaterials with periodic and nonperiodic styles. The applications of ENZ metamaterials are discussed in Sections 4 and 5, from the perspectives of microwave engineering, optics, and quantum physics. The authors conclude in Section 6 by presenting an outlook on the development of ENZ metamaterials and discussing the key challenges addressed in future works.

Keywords: epsilon-near-zero, metamaterials, antennas, integrated circuits, plasmonics

ISBNs: 9781009124416 (PB), 9781009128339 (OC)
ISSNs: 2399-7486 (online), 2514-3875 (print)

Contents

1 Introduction

1.1 Family of Near-Zero-Index Materials/Metamaterials

The foundation of electromagnetic theory (1865) by James Clerk Maxwell and the pioneering experiment (1887) by Heinrich Hertz, exciting moments in physics, eventually led to the discovery of the electromagnetic wave. In the past one and a half centuries, the electromagnetic wave has played an essential role in wireless communication, target detection, remote sensing, medical engineering, and other fields, all of which substantially promote the prosperity of modern society. The past century has also witnessed an increasing interest in devices that provide better control of the transmission, scattering, and radiation of electromagnetic waves. Among the well-known examples are microwave waveguides, optical fibers, lenses and mirrors, and radio-frequency (RF) antennas and arrays, to name just a few. On the other hand, advances in material science and manufacturing crafts have opened exciting opportunities to conceive desired materials and structures at the level of microns and even nanometers. Assisted by elaborately designed materials, we are capable of sculpting wave–matter interactions in the deep subwavelength scale.

Generally, the macroscopic response of a material to the electromagnetic wave is depicted by its constitutive parameters: permittivity ε and permeability μ, presenting the electric and magnetic responses when the material is impinging on the electromagnetic wave. Another important material parameter relating to those two fundamental quantities is the optical refractive index $n = (\varepsilon\mu)^{1/2}$. Figure 1.1 illustrates the classification of various types of materials on the ε-μ plane, in which there are four quadrants with four different combinations of ε and μ. The free space has a permittivity ε_0 and a permeability μ_0, and the relative permittivity and permeability of a certain material are therefore defined as $\varepsilon_r = \varepsilon/\varepsilon_0$ and $\mu_r = \mu/\mu_0$, respectively. Most naturally occurring materials, such as dielectrics, feature a permeability and a permittivity larger than zero and hence are termed double positive materials, falling under the first quadrant of Fig. 1.1. As classified in the second quadrant, the materials with $\varepsilon < 0$ while $\mu > 0$ are designated as epsilon-negative materials. Many electric plasmonic materials behave in this manner below their plasma frequencies, such as the noble metals in the visible frequencies. The materials with $\varepsilon > 0$ while $\mu < 0$, categorized in the fourth quadrant, are called mu-negative materials. Some magnetic gyrotropic materials exhibit this characteristic in certain frequency regions. The double negative materials, lying in the third quadrant, also known as left-handed materials and negative refractive index (NRI) materials, were first proposed theoretically by V. G. Veselago in 1960s, but did not attract great interest for the next 30 years until their realization based on the emergent artificially structured materials (i.e., metamaterials [1–8]).

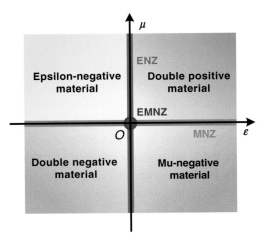

Figure 1.1 Classification of materials on the ε–μ plane.

Generally, the concept of metamaterial refers to a macroscopic composite of periodic or nonperiodic subwavelength resonant or nonresonant structures, which can be described by the effective constitutive parameters of a virtually homogenous material not existing in nature, such as double negative materials [1, 2]. The electromagnetic responses outside the materials are identical between the periodic metamaterial and the virtually homogenous material. The concept of metamaterials is widely studied and has performed well from microwave [3] to optical regions [4], and from general theory [5] to engineering applications [6]. Via suitable design of the unit cells and lattice configurations, metamaterials can be engineered to exhibit effective double positive, epsilon-negative, mu-negative, or double negative responses, promoting the development of metamaterials or meta-structures, as well as the emergence of various meta-devices [7]. Then, the three-dimensional bulky metamaterials evolve toward two-dimensional flatland structures (i.e., metasurfaces) for purposes of easy fabrication and advanced control of wave transmission and reflection [8].

The metamaterials falling under the transitional regions (shown as red and blue strips in Fig. 1.1) between two quadrants with refractive index near zero exhibit counterintuitive but interesting near-zero responses, such as optical nonlinearity enhancement, wave supercoupling, and directive emission [9, 10, 11]. Such a metamaterial is therefore called near-zero-index (NZI) metamaterial. In addition to the isotropic NZI metamaterials, anisotropic NZI metamaterials are also being proposed and investigated, whose constitutive parameter ε or μ is described by a tensor with one or several diagonal components close to zero [12]. Furthermore, the concept of NZI metamaterials can be readily extended to other fields of physics, such as acoustics [13] and thermology [14], exhibiting excellent control ability for various types of waves. For example, the permittivity ε and permeability μ can be mapped onto the reciprocal of bulk modulus $1/\kappa$ and mass density ρ, respectively, in the acoustic wave equation. In this manner, it is possible to analyze acoustic NZI metamaterials in analogy with their electromagnetic counterparts. Concretely, depending on one or both constitutive parameters' approach to zero, NZI metamaterials can further be categorized as epsilon-near-zero ($\varepsilon \approx 0$, ENZ), mu-near-zero ($\mu \approx 0$, MNZ), and epsilon-and-mu-near-zero ($\varepsilon \approx 0$ and $\mu \approx 0$, EMNZ) metamaterials. The main difference among these three types of NZI metamaterials is the intrinsic impedance $Z = (\mu/\varepsilon)^{1/2}$, which is near zero for MNZ, near infinity for ENZ, and a regular complex number for EMNZ. The magnetic field is irrotational for ENZ, and the electric field is irrotational for MNZ. For EMNZ, both fields are irrotational.

In this Element, we focus on the ENZ metamaterials, performing a systematic review from the aspects of general concept, intriguing phenomena, and engineering applications. Corresponding to a particular zone in the constitutive parameter space, ENZ metamaterials are expected to yield different wave phenomena. We have known that the double positive and double negative materials or metamaterials are electromagnetically transparent, as the wave numbers in such media are real and the traveling wave state is supported. It can be understood that the single-negative materials/metamaterials are electromagnetically opaque due to the waves becoming evanescent in their bodies. By a simple inspection of the ε-μ plane in Fig. 1.1, one may raise a natural question: How would the electromagnetic waves propagate in ENZ metamaterials, which are situated in the critical zones? In fact, electromagnetic waves in the ENZ metamaterial exhibit a spatially static behavior, which is essentially different from that in other metamaterials. This question is addressed in Section 2 of this Element, and rich ENZ phenomena and functions are demonstrated and analyzed in the rest of the Element. Even though the natural ENZ materials (i.e., plasmonic materials) exist in the optical domain, nonetheless their intrinsic loss hinders the development of ENZ applications [15]. The idea of ENZ metamaterials using artificial structures to imitate ENZ behavior presents potential in practical applications. The key feature of this Element is presenting various engineering applications based on the concept of ENZ metamaterials.

1.2 History of Epsilon-Near-Zero (ENZ) Metamaterials

The pioneering work to analyze the wave dynamics in ENZ metamaterials may date back to 2004 [16]. In this work, Ziolkowski studied analytically and numerically the wave propagation and scattering in the ENZ medium and revealed that the electromagnetic fields in the ENZ medium take on a static characteristic in space, yet remain temporally dynamic. The theory can also be adopted in other types of NZI medium with different impedance matching considerations. Arguably the first mind-bending phenomenon predicted in the ENZ material was that, as Silveirinha and Engheta theoretically demonstrated in 2006, the electromagnetic waves could be squeezed into a narrow two-dimensional ENZ channel and achieve a total transmission with near-zero reflection, an effect referred to as "supercoupling" [17]. Remarkably, the super-coupling effect depends neither on the length nor on the shape of the narrow ENZ channel, exhibiting substantial differences from the classic Fabry–Perot resonant transmission. In this work, the area of the ENZ channel should be infinitely small to achieve the total transmission property, which is also achieved by using $\mu \approx 0$ (i.e., EMNZ medium matching in this case). Subsequently in 2008, the supercoupling effect was experimentally validated using an extremely narrow waveguide operating at the cutoff frequency [18] and a metamaterial composed of planar complementary split-ring resonators inside a waveguide [19]. Both methods use a waveguide to imitate ENZ behavior but in different ways. In [19], a typical periodic metamaterial para-digm is used to achieve an ENZ effect. However, for [18], the waveguide operates at its cutoff frequency and can be treated as a new type of nonperiodic method to achieve an ENZ effect, which is further investigated in Section 2. In its early stage, interest in the ENZ metamaterials was strongly triggered by their potential to shape the radiation pattern and control the wave front. As discussed in Enoch's work (2002) [20], the stretched wavelength of electromagnetic waves in a low-index medium could provide the phase and magnitude uniform-ity of fields over an electrically large aperture and benefit the generation of a directive radiation for the broadside beam.

Subsequently, the roles that ENZ metamaterials play in wave–matter inter-actions were actively investigated and proved by experiment. In 2010, Ciattoni et al. theoretically analyzed the extremely nonlinear electrodynamics in an ENZ metamaterial that features a vanishing linear dielectric permittivity [21]. This work opened new horizons to boost the optical nonlinearity assisted by ENZ response, and a series of experiments was performed in nonlinear enhancement later. For example, Suchowski et al. in 2013 experimentally validated the phase-mismatch-free, four-wave-mixing process based on an ENZ metamaterial [22]. Besides enhancing materials' nonlinearity, ENZ metamaterials also have the potential to boost the extremely weak optical nonlocality. In Pollard's work (2009), the ENZ metamaterial is proved to be a desired platform for observing the spatial dispersion and additional wave, which do not exist in the local materials [23]. Another intriguing aspect of the wave–matter interaction in the

ENZ metamaterial is associated with the optical nonreciprocity. In 2013, Davoyan theoretically demonstrated that the magnetized ENZ metamaterial is a promising candidate to achieve one-way transmission, even with magnetically switched transparency and opacity [24]. ENZ metamaterials were also introduced into the regime of quantum optics, exhibiting functionalities in collective superradiance enhancement (2013) [25].

Another important timeline is for the development of the implementation schemes of ENZ metamaterials. As early as 1962, Rotman originally proposed to emulate plasmonic materials with the use of parallel-plate waveguides operating at the TE_{10} mode [26]. Specifically, the waveguide at the cutoff frequency of TE_{10} can mimic the behavior of ENZ material. Following this idea, in 2008 Edwards et al. fabricated waveguide ENZ metamaterials to verify the supercoupling effect in microwave frequency [18]. In the context of periodic artificial structures, ENZ metamaterials can also be equivalently realized via the zeroth-order mode of the left-handed transmission lines [27] and stacked alternative positive and negative dielectric layers [28]. In 2011, Huang et al. [29] showed that around the Dirac cone at the center of the Brillouin zone, the dielectric photonic crystal exhibits the desired zero-index property. Then it followed the work by Li et al. to realize the on-chip zero-index metamaterial (2015) [30] operating at the infrared frequency based on periodic structures. To tailor the macroscopic scattering properties of the ENZ metamaterial, Inigo et al. proposed the concept of photonic doping (2017) [31], employing dielectric impurities to control the effective permeability of the ENZ metamaterial. In 2019, Zhou et al. [32] introduced the substrate-integrated waveguide to realize the ENZ metamaterial and accommodated the concept of photonic doping in a planar architecture compatible with integrated circuits processes.

A handful of subfields in physics and engineering are rising, inspired by the ENZ metamaterials. An exciting subject is emergent optical metamaterial circuitries, which are discussed in detail in this Element. In 2005, Engheta et al. [33] demonstrated that deeply subwavelength non-plasmonic or plasmonic nanoparticles can serve as lump elements with effective capacitance or inductance, operating as nanocapacitors and nanoinductors for nanooptics. Subsequently in 2007, the concept of "metatronics" [34] was put forward to represent a class of metamaterial-inspired nanocircuits based on the scientific contribution in [33]. To connect those optical lump elements, the optical displacement-current conduit was proposed in 2009 by Alù et al. [35], who harnessed the ENZ metamaterial to efficiently confine and route the displacement current, analogous to the conduction current in DC circuits. Over the years, metatronics has been actively applied to design optical circuit modules; for example, optical filters with subwavelength scales [36]. In an important step toward further progress, Li et al. in 2016 [37] explored the platform of the waveguide to implement the idea of metatronics, with much improved robustness and reduced optical losses from natural plasmonic materials [38, 39, 40].

1.3 Outline of the Element

This Element is dedicated to introducing the unusual wave phenomena arising from an extremely small optical index of refraction, and to shedding light on the underlying mechanisms, with the primary focus being on the basic concepts and fundamental wave physics. We reveal the exciting applications of ENZ metamaterials, which have profound impacts over a wide range of fields of science and technology. The sections are organized as follows. In Section 2, we demonstrate the extraordinary wave properties in ENZ metamaterials, analyzing the unique wave dynamics and the resulting effects. Section 3 introduces various realization methods of the ENZ metamaterials with periodic and non-periodic styles. The applications of ENZ metamaterials are discussed in Sections 4 and 5 from the perspectives of microwave engineering, optics, and quantum physics. We close in Section 6 by presenting an outlook on the development of ENZ metamaterials and discussing the key challenges to be addressed in future research.

2 Wave Properties in ENZ Metamaterials

2.1 Wave Dynamics at the ENZ Condition

In this section, we discuss the wave properties of ENZ metamaterials from the general electromagnetic theory and derive the counterintuitive characteristics of wave–matter interactions, including the basic wave dynamics, space and time decoupling, and wave supercoupling. Here, we start the discussion of the wave dynamics in the ENZ condition by firstly inspecting the basic equations in electromagnetics. Consider the case that a monochromatic plane wave propagates in an isotropic and homogenous medium and that the temporal period (i.e., frequency f) and the spatial period (i.e., wavelength λ) of the propagating wave are related via a simple expression:

$$f \cdot \lambda = c / \sqrt{\varepsilon_r \mu_r}, \tag{2.1}$$

where c is the speed of light in a vacuum, while ε_r and μ_r denote the relative permittivity and permeability of the medium. The right side of Eq. (2.1) is the phase velocity of light inside the medium. Consider that as ε_r approaches zero (i.e., ENZ condition), the phase velocity of light diverges to infinity. Under a given frequency f, it is also readily seen from Eq. (2.1) that the wavelength in the medium should approach infinity as the relative permittivity ε_r vanishes to zero. Here the general conclusion is that when ε_r decreases to near zero, for different frequencies of the wave, the wavelength goes to infinity, the phase velocity goes to infinity, but the velocity of power flow goes to near zero, exhibiting static-like behavior inside the ENZ medium. Therefore, the diverging phase velocity and extremely stretched wavelength at the ENZ condition imply a suppressed variation of the electromagnetic fields in phase and amplitude over a large-scale space. Importantly, such a spatially static field configuration can be allowed at a nonzero frequency, which is a unique wave property in ENZ and other NZI materials.

To proceed, we check the ENZ limit ($\varepsilon_r \approx 0$) from the behavior of Maxwell's equations, in which two curl equations (i.e., Maxwell–Ampère law and Faraday's law) respectively read as follows:

$$\nabla \times \mathbf{H} = \frac{\partial \mathbf{D}}{\partial t} + \mathbf{J} \tag{2.2}$$

$$\nabla \times \mathbf{E} = -\frac{\partial \mathbf{B}}{\partial t}, \tag{2.3}$$

where $\mathbf{E}$ and $\mathbf{H}$ are respectively the electric field and the magnetic field; $\mathbf{D} = \varepsilon_r \varepsilon_0 \mathbf{E}$ and $\mathbf{B} = \mu_r \mu_0 \mathbf{H}$ denote respectively the electric and magnetic flux densities. We assume that the area of interest is source-free (i.e., the current density $\mathbf{J} = 0$). At the ENZ condition ($\varepsilon_r \approx 0$), the electric flux density $\mathbf{D}$ vanishes, and therefore the curl of the magnetic field is zero, as seen from Eq. (2.2). Consequently, at the ENZ condition, the temporally varying magnetic field $\mathbf{H}$ is distributed in space the same way that the static magnetic field is. In particular, we consider a transverse-magnetic (TM) case where the magnetic field is polarized along one axis (e.g., the z-axis) and invariant along that axis, that is, $\partial_z \mathbf{H} = 0$. Substituting $H = H(x,y)\hat{z}$ into Eq. (2.2) and imposing the source-free and ENZ condition, we obtain $\nabla \times (H\,(x,y)\hat{z})$, which reduces to $H(x,y) \approx 0$. As seen, for this TM wave (oriented with respect to the z-axis) in the ENZ material, the magnetic field is a constant in any x-y cut plane. The spatially homogenous magnetic field distribution is a salient feature in two-dimensional (2D) ENZ materials or metamaterials, which is related to numerous unusual phenomena and functionalities. One of the most straightforward examples is the wave front manipulation. In the case of TM excitation, due to continuity of the tangential magnetic field, the wave front of the scattered or radiated magnetic field from an ENZ body should be conformal to its surface.

ENZ materials/metamaterials are also characterized by a diverging intrinsic impedance $\eta = \sqrt{\mu/\varepsilon}$, which results in the impedance mismatch with free space. As another general conclusion, when ε_r decreases to near zero, the intrinsic impedance of a wave inside the ENZ medium goes to infinity, presenting a value similar to that of a perfect magnetic conductor (PMC).

It is worth noting here that the infinite phase velocity at the ENZ condition does not contradict the causality principle in special relativity, because the phase velocity does not correspond to the velocity of either information or energy. What matters in the causality principle is the group velocity of the wave, which should be smaller than c. The group velocity in a dispersive homogeneous medium is defined and given by

$$v_g = \frac{\partial \omega}{\partial k} = \frac{2c}{2\sqrt{\mu_r \varepsilon_r} + \omega\sqrt{\frac{\mu_r}{\varepsilon_r}\frac{\partial \varepsilon_r}{\partial \omega}} + \omega\sqrt{\frac{\varepsilon_r}{\mu_r}\frac{\partial \mu_r}{\partial \omega}}}, \tag{2.4}$$

where $\omega = 2\pi f$ is the angular frequency and $k = 2\pi/\lambda$ is the wave number or propagation constant in the medium. In a lossless ENZ medium, the group velocity, evaluated by Eq. (2.4) after inserting $\mu_r = 1$, tends to zero as the slope of the permittivity as a function of angular frequency is finite. As discussed in [29], the NZI metamaterial with a single parameter (permittivity or permeability) approaching zero realized via a photonic crystal can feature a quadratic dispersion, $\delta\omega = 0 + O(\delta k^2)$, that is almost flat near $\mathbf{k} = 0$. As a case study, we consider a plasmonic material whose permittivity is described by the Drude-type dispersion: $\varepsilon_r(\omega) = 1 - \omega_p^2/(\omega^2 + i\omega\omega_c)$, where ω_p and ω_c are respectively the plasma frequency and the collision frequency. It is clear that in the lossless limit ($\omega_c \to 0$), the permittivity ε_r vanishes and the phase velocity of the wave diverges at plasma frequency ω_p, while the group velocity $v_g(\omega_p) = 2c\sqrt{\varepsilon_r(\omega_p)}/[\varepsilon_r(\omega_p) + 1]$ approaches zero. In fact, when the wave is excited in the ENZ medium, it requires a certain transitory time to build up the steady state of the spatially static field configuration, which ensures that the ENZ materials/metamaterials abide by the causality principle. On the other hand, for the practical system, the existing loss increases the group velocity inside the ENZ medium and also increases the transitory time needed to build up the spatially static field. For a regular ENZ medium with a certain amount of loss, the significantly reduced group velocity can be equivalently interpreted as the slowing down of light, which substantially increases the time scale of wave–matter interaction.

On the other hand, for the EMNZ metamaterials, the related intrinsic impedance $\eta = \sqrt{\mu/\varepsilon}$ is finite, since both the permeability μ and the permittivity ε

approach zero simultaneously, and so they can be interpreted as a kind of "matched" ENZ metamaterials [41]. The EMNZ medium is also known for its linear dispersion [29] and finite group velocity of light. By letting μ_r and ε_r approach zero while μ_r/ε_r and ε_r/μ_r are kept finite, it is easy to check in Eq. (2.4) that the group velocity v_g in a dispersive EMNZ medium is finite. In the EMNZ metamaterial realized via a photonic crystal with accidental degeneracy at $\mathbf{k} = 0$, a linear dispersion relationship of light can be obtained as $\delta\omega = v_g\,\delta k + o(\delta k)$ [29]. Recall the constitutive relationship $\mathbf{B} = \mu\mathbf{H}$ and that by imposing $\mu \approx 0$, Eq. (2.3) reduces to $\nabla \times \mathbf{E} \approx 0$, which is actually a high-frequency effective implementation of Kirchhoff's voltage law in a differential form. Evidently, in the EMNZ metamaterials, both the electric field and the magnetic field are spatially static, and furthermore, they no longer couple with each other. As systematically discussed in [42], a peculiar wave–matter interaction appears with "opening up" and "stretching" the space, as well as behaving electromagnetically as a single point even though the actual volume is electrically large. In this manner, the behavior of time-harmonic fields in an infinite EMNZ metamaterial can be fully analogous to that in electro- or magnetostatic fields. The electromagnetic wave with time convention $\exp(-i\omega t)$ suggests the exciting possibility of "DC optical circuits" (i.e., circuits operating at an optical frequency with a distributed wave characteristic), but the displacement current can be modeled as that in lumped circuits.

2.2 Decoupling of Spatial and Temporal Behavior of Waves

In this section, we pay specific attention to the property of decoupled temporal and spatial variation of electromagnetic fields in the ENZ medium and shed light on its underlying implications. Generally, as manifested in the wave equation,

$$\nabla^2 \begin{Bmatrix} \mathbf{E} \\ \mathbf{H} \end{Bmatrix} - \varepsilon\mu\frac{\partial^2}{\partial t^2} \begin{Bmatrix} \mathbf{E} \\ \mathbf{H} \end{Bmatrix} = 0, \tag{2.5}$$

the space-domain derivative operator ∇ and the time-domain derivative operator $\partial/\partial t$ are coexisting when describing a dynamic field behavior. The resulting effect is the coupled temporal and spatial variations of the electromagnetic fields. In the ENZ limit ($\varepsilon \approx 0$), however, the term associated with the time derivative ($\partial/\partial t$) in Eq. (2.2) vanishes to near zero, leading to the decoupling of temporal dynamic and spatial variation of the electromagnetic wave. This result arises from $\varepsilon \approx 0$, but it also can be effectively achieved when the time derivative is near zero, presenting static field behavior inside the ENZ medium. Such a result is also feasible for metamaterials with MNZ or EMNZ responses. To

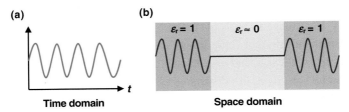

Figure 2.1 Decoupling of temporal and spatial variation of an electromagnetic field. Temporally oscillating field (a) with suppressed spatial variation in the ENZ region (b)

gain an intuitive understanding of this point, we consider a plane wave oscillating periodically in the time domain (illustrated in Fig. 2.1(a)) that transmits through an air–ENZ–air structure. A snapshot of the spatial distribution of the wave is schematically presented in Fig. 2.1(b). Although the wave remains temporally dynamic, it exhibits a spatially static behavior – uniform phase and amplitude distributions – in the ENZ region.

Due to the associated temporal and spatial properties of the electromagnetic wave, which are commonly observed in conventional materials and metamaterials, the operation frequencies of microwave and optical devices are usually related to their physical dimensions and detailed geometries. For example, the dipole antenna radiates efficiently when its physical length is about half the wavelength. The Fabry–Perot laser operates when the dimension of the cavity meets the resonance conditions. In this regard, ENZ metamaterials, which ratify the decoupled temporal and spatial wave properties, offer an exciting opportunity to conceive and realize geometry-irrelevant microwave/optical devices. As a representative example, let us consider geometry-irrelevant cavities made of ENZ metamaterials [43]. As conceptually illustrated in Fig. 2.2, three different cavities with arbitrary shapes are constructed with the 2D ENZ host as the background and a dielectric particle with cross-sectional area A_i. The ENZ hosts have different cross-sectional shapes but the same area, A_h. All the cavities are bounded by perfect electric conductor (PEC) walls. The resonance frequency of the cavity is formally defined as the eigenvalue of the source-free, time-harmonic wave equation subject to the PEC boundary condition ($\mathbf{n} \times \mathbf{E} = 0$). For this problem, we apply Faraday's law on the boundary of the ENZ host to yield [43]:

$$\omega = \frac{i}{\mu_0 A_h} \frac{1}{H_h} \oint_{\partial A_i} \mathbf{E} \cdot \mathbf{dl}, \tag{2.6}$$

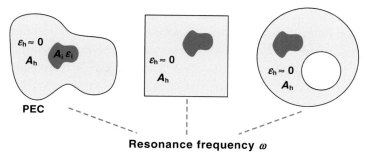

Figure 2.2 Conceptual sketch of geometry-irrelevant resonators [43], the 2D cavities composed of 2D ENZ host with cross-sectional area A_h and a 2D dielectric particle with cross-sectional area A_i. The cavities are bounded by PEC walls.

where μ_0 is the permeability in a vacuum, and H_h represents the magnetic field in the ENZ host, which has been proved to be uniformly distributed. We have used the fact that the tangential electric field is zero on the PEC walls. The circulation of the electric field along the periphery of A_i over the background magnetic field H_h reflects the surface impedance of the particle embedded in the ENZ host. It is clearly seen from Eq. (2.6) that the resonance frequency of the cavity is completely determined by the area of the ENZ host A_h and the properties of the inner dielectric particle, encapsulated in the surface impedance term. Hence, it is clear that the resonance frequency is invariant under the geometry transformation and has no dependence on the shape of the external boundary of the cavity. Additionally, the existence of the resonant mode is independent of the topology of the cavity. As shown on the right side of Fig. 2.2, by introducing a 2D hole to the ENZ host, we transform the simply connected cavity into a multiply connected one, while the resonance frequency of the cavity is immune to change if the overall area of the ENZ host is kept the same.

The ENZ-host cavities are completely bounded by the PEC walls. In fact, the concept of the geometry-irrelevant cavities can be extended further to the case that the cavities are interacted with the free space. From the intrinsic impedance point of view, the impedance of ENZ host is infinite, and that for free space is 377 ohms. If an observer inside the ENZ host looks toward the free space, the reflection coefficient is approaching −1, which is identical to that of a PEC. Therefore, even though we use an open boundary instead of a PEC wall, the magnetic field still remains uniform. The open boundary enables a class of geometry-irrelevant antennas, whose geometry and even radiating aperture distribution can be flexibly designed with no influence on the operation frequency. The functionalities of geometry-irrelevant antennas made of ENZ metamaterials are discussed in detail in Section 4. Another important example of a geometry-irrelevant and deformable device is the ENZ-metamaterial transmission line, which features zero phase delay and a high transmission rate immune to arbitrary bending or stretching. The physics underlying such a geometry-irrelevant transmission line is associated with the unusual supercoupling effect occurring at the ENZ condition, which is described and analyzed in Section 2.3. In essence, the geometry independence of the ENZ metamaterials stems from a spatially static field configuration with an effectively stretched wavelength, which in turn reduces significantly the influence of the geometry.

2.3 Supercoupling of Waves

When the propagating wave in an open space encounters the inhomogeneity of the background, it is scattered in various directions. Similarly, the guided wave, transmitted in a microwave waveguide or an optical fiber, for example, is expected to be partially reflected back when it encounters obstacles or experiences the geometrical deformation of the transmission lines. In general, the amount of reflection increases when the structural deformation in the guiding structure becomes more significant. However, the rise of ENZ metamaterials provides an exciting possibility of "squeezing" the wave into an arbitrarily deformed narrow channel and achieving a perfect transmission. Such an anomalous tunneling of the wave is called "supercoupling" [17, 18, 19]. At first glance, one might associate this term with the tunneling effect in quantum mechanics, where a microscopic particle has a chance to overcome a barrier even with a potential higher than its kinetic energy. It is interesting to compare these two effects, as in both cases the transfer of energy or substance is achieved in seemingly impossible conditions. However, the underlying physics can be quite different. The quantum tunneling is allowed due to the nonzero probability wave of the particle on the other side of the barrier, while the supercoupling effect is essentially a result of suppressed total magnetic flux in the narrow ENZ channel and the matched impedance. Besides quantum perspective, the supercoupling effect is achieved based on special resonant modes, such as localized plasmonic resonance, used to pass the light through subwavelength holes [44], and periodically arranged slabs with alternating positive and negative permittivity (i.e., equivalent inductor–capacitor resonance [45]). From this point of view, the ENZ supercoupling behaves as a zeroth-order mode to couple the wave from one side to the other side.

The general phenomenon of wave "supercoupling" was first proposed in [17], experimentally verified in [18], and theoretically analyzed using both wave

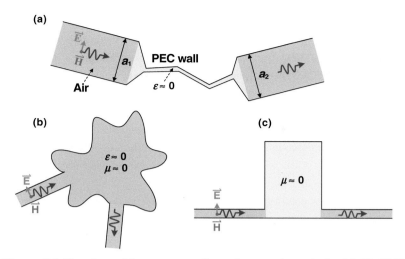

Figure 2.3 Sketches of the supercoupling of waves through the (a) 2D ENZ channel [17], (b) 2D EMNZ channel [31], and (c) 2D MNZ channel [48]. For all the cases, the incident wave is polarized with the magnetic field along the out-of-plane axis, and the input and output waveguides are filled with air. All the structures except for those at the ports are surrounded by PEC walls.

propagation theory [46] and transmission line theory [47]. A conceptual sketch is presented in Fig. 2.3(a) to illustrate the supercoupling phenomenon through an irregular-shaped narrow channel filled with the ENZ material. The 2D structure is assumed to be invariant along the out-of-plane axis. The heights of the air-filled input and output waveguides are a_1 and a_2, respectively. The incident wave is polarized with the magnetic field along the out-of-plane axis and the electric field on the plane. Under the condition that the material filling the channel has an extremely low permittivity (i.e., near zero), the reflection coefficient R and the transmission coefficient T of the magnetic field have been derived analytically:

$$R = \frac{(a_1 - a_2) + ik_0\mu_{r,c}A_c}{(a_1 + a_2) - ik_0\mu_{r,c}A_c} \qquad (2.7)$$

and

$$T = \frac{2a_1}{(a_1 + a_2) - ik_0\mu_{r,c}A_c}, \qquad (2.8)$$

where k_0 is the wave number of free space, $\mu_{r,c}$ denotes the relative permeability of the ENZ metamaterial inside the narrow channel, and A_c is the cross-sectional area of the ENZ channel. Note that the relation $R + 1 = T$ holds, based on the tangential electric and magnetic fields' continuity at the medium boundary. The procedure used to yield Eqs. (2.7) and (2.8) applies Faraday's law on the boundary of the ENZ channel and uses the magnetic field uniformity in the ENZ region. As clearly seen from Eqs. (2.7) and (2.8), if we would like to achieve perfect tunneling effect ($|R| = 0$, $|T| = 1$), two conditions have to be satisfied simultaneously: (i) $a_1 = a_2$ and (ii) $k_0 \mu_{r,c} A_c \rightarrow 0$. It is observed from Eq. (2.8) that, at the supercoupling, the phase of transmission coefficient is exactly zero (i.e., $\text{Arg}[T] = 0$), which agrees with our expectation concerning the behavior of NZI materials.

Next, we discuss conditions (i) and (ii) of the supercoupling effect. Because the magnetic fields are equal on the interfaces of the ENZ region with input and output waveguides, the power flow densities in the perfect tunneling situation, identical at the input and output interfaces, are $\eta_0 |H_c|^2/2$ (where η_0 is the wave impedance in free space and H_c is the magnetic field in the ENZ region). Hence, condition (i) states the requirement of the equal incident and output powers in the supercoupling situation. In condition (ii), the term $\mu_{r,c} A_c$ is proportionate to the total flux of the magnetic field in the ENZ region. Therefore, under a given wave number k_0, condition (ii) implies the extremely weakened magnetic response of ENZ channels. The first possibility to satisfy condition (ii) is to let $A_c \rightarrow 0$, namely, the cross-sectional area of the ENZ channel should be infinitesimal compared with the scale of the operation wavelength. That is the case in Fig. 2.3(a), where the ENZ channel is extremely narrow. Notably, as no restrictions are imposed on the detailed shape of the 2D ENZ channel, we can arbitrarily deform or distort the channel while having no influence on the performance of the supercoupling. Another possibility to satisfy condition (ii) is to let $\mu_{r,c} \rightarrow 0$, that is, the channel is filled with the EMNZ metamaterial, which can be treated as the matched ENZ metamaterial. In this case, the requirement for a vanishing cross-sectional area of the channel can be substantially loosened; therefore, the EMNZ supercoupling effect is allowed in the channel with arbitrary shapes, sizes, and even complex topologies, as conceptually illustrated in Fig. 2.3(b). Several intriguing concepts associated with EMNZ supercoupling are demonstrated in the literature [42], among which are "opening up the space" and achieving the "electromagnetic point" for any volume, even though it is electrically large. Suppose one splits an air-filled parallel-plate waveguide into two sections and links them by an arbitrarily shaped and sized EMNZ channel; according to the property of EMNZ supercoupling, external electromagnetic properties such as transmission amplitude and phase of the waveguide won't be affected at all. That implies we could have a space with large bounded physical volume while the electromagnetic wave hardly perceives it. The concept of electromagnetic point can be interpreted from the extremely stretched wavelength (i.e., infinitely large wavelength), which suppresses the spatial variation of fields and therefore effectively shrinks the space to a point.

We have shed light on the supercoupling effect from the perspective of diminished magnetic response (i.e., the magnetic flux along the ENZ channel), which ensures the continuity of the electric fields on the interfaces of the channel with the input and output waveguides. On the other hand, one can understand this effect more simply from the perspective of impedance matching [47]. It is noted that the characterizing impedance of a parallel-plate waveguide is proportionate to the wave impedance $\eta = \sqrt{\mu/\varepsilon}$ in the medium and the distance between the two PEC plates, namely its height. Resultantly, the dramatically increased wave impedance due to the near-zero permittivity can be compensated by structurally narrowing the channel, which corresponds to the case of ENZ supercoupling shown in Fig. 2.3(a). Alternatively, we can avoid the explosion in wave impedance by letting the permeability μ go to zero simultaneously [48, 49], which explains the EMNZ supercoupling shown in Fig. 2.3(b). Similarly, it is easy to understand the MNZ supercoupling effect, as shown in Fig. 2.3(c), where increasing the height of the channel remedies the extremely reduced wave impedance due to the vanishing permeability μ and thus realizes a matched impedance with the outside waveguides. In the work of [48], the "supercoupling" channel has a much larger cross-sectional area than do the input and output waveguides, totally opposite to the setup of ENZ super-coupling. The same concept is also realized in [49] by inserting periodical resonators inside the waveguide, providing another effective path to the general supercoupling of MNZ. It also can be understood from the impedance matching point of view.

Next, we present and analyze several important implications of the ENZ supercoupling effect, performing useful applications based on the ENZ behavior. It has been demonstrated in [50] that the ENZ supercoupling effect can enable the guidance of light through arbitrarily curved routes, and this finding has been experimentally verified in the microwave region [51]. The ENZ supercoupling behavior is maintained with different bending angles of the channels with high quality factor, but its operation bandwidth can be an issue. Since the first theoretical prediction (2006) [17] and the first experimental demonstrations (2008) [18] of the ENZ supercoupling effect, it has been known as an extremely narrowband phenomenon (i.e., the ENZ supercoupling exists only within a small frequency range). The reasons are twofold: First, the realistic ENZ metamaterials or plasmonic materials are dispersive, whose permittivity functions cross zero at certain frequencies. Second, the impedance matching in such an extreme-index condition can be highly sensitive, which further shrinks the bandwidth of the tunneling. On the other hand, the narrow-band, or the high-quality-factor property of the supercoupling effect makes it fertile ground to facilitate wave–matter interactions. The narrow fractional bandwidth can be translated into an enhanced quality factor, which is generally defined as the energy stored per cycle over that which is dissipated. Hence the ENZ supercoupling effect is related to the high energy density in the channel, that is, the enhanced electromagnetic fields.

For engineering applications, wide bandwidth is important to support higher data transmission rate and spectrum coverage of different services. As a feasible solution, multiple ENZ tunnels are designed in [52] by organiz-ing multiple narrowband transmissions at different frequencies. As another effective method to solve the narrowband issue, the reconfigurable technique, also known as the tunable strategy, is utilized in the ENZ supercoupling operating range enlargement [53, 54, 55]. In order to dynamically tune the operating frequency of the ENZ supercoupling, slots are etched on the waveguide to disturb the current distributions along the waveguide. Tunable elements or components, such as PIN diodes or varactors, are soldered onto the slots to control the state of the current temporally. In this way, the frequency of ENZ supercoupling is tuned within a wide frequency region, with a wide frequency tuning range of over 10 percent [54]. Last but not least, another way to enhance the supercoupling bandwidth was studied in [56]. A method of optical conformal matching is proposed (e.g., using regular dielectric [$\varepsilon_r = 10$] instead of epsilon-near-zero substrate inside the narrow channel, or using epsilon-large dielectric [$\varepsilon_r = 10,000$] instead of air inside the input and output waveguides). In this way, all the features of ENZ super-coupling are maintained, but with enhanced bandwidth and lower dielectric loss. The key to this method is using epsilon-large dielectric, which proposes another challenge to materials science. With the merits of ENZ metamater-ials, the supercoupling effect has been expanded from the optical region to the microwave region, following the exact same design rules and performance. For example, the light can be funneled through a subwavelength aperture [57]; the surface wave (e.g., surface plasma polariton [SPP]) is also tunneled without the considerations of channel bending [58]. This function is also useful in microwave systems, such as spoof surface plasma polariton (SSPP) guiding [59] and coaxial-to-waveguide transition [60], exhibiting new methods to design microwave components or circuits.

Note that the energy density of the electric field is not close to zero at the ENZ metamaterials. The time-averaging electromagnetic energy density in a homogenous dispersive medium should be computed by the following formula:

$$\overline{W} = \frac{1}{4}\frac{\partial(\omega\varepsilon)}{\partial\omega}|E|^2 + \frac{1}{4}\frac{\partial(\omega\mu)}{\partial\omega}|H|^2, \tag{2.9}$$

where E and H represent the amplitudes of the electric field and the magnetic field, respectively. As the amplitude of the magnetic field in ENZ channel is equal to that of incidence at the supercoupling condition, the electric field in the ENZ channel is dramatically enhanced compared with the incidence. Such a field enhancement effect can be independent of the position in the ENZ channel and becomes more significant as we further narrow the channel. Consequently, the platform of ENZ channels can be harnessed to raise the sensitivity of dielectric sensors, boost the weak optical nonlinear effects, and even enhance the collective emission of the quantum emitters. The detailed applications related to the ENZ metamaterials and the supercoupling effect are discussed in Section 4 of this Element.

3 Realizations of ENZ Metamaterials

3.1 Plasmonic Materials

This section presents an overview of the realizations of ENZ metamaterials by using different methodologies, including both natural and artificial ways. As the first feasible path to achieving ENZ parameters, the plasmonic materials are systematically studied at their plasma frequency. Plasmonic materials, known as noble metals in optical frequencies, produce a special dispersion of permittivity with the positive, zero, or negative real part of permittivity in a certain frequency regime. The plasmonic materials have become an intense research subject over the past decades, owing to their unprecedented capability of concentrating and manipulating electromagnetic fields and waves on a subwavelength scale. Such plasmonic materials enable the combination of photonics and electronics using nanostructures, leading to the oscillating mode known as plasmon or surface plasmon. Particularly, surface plasmon polaritons (SPP) can be supported along the metal's surface, enhancing light–matter interactions and leading to many near-field applications such as surface-enhanced Raman spectroscopes, biosensors, and so on [61]. To date, plasmonic materials have been widely employed in various nanophotonic devices such as subwavelength waveguides, optical nanoantennas, hyperlenses, optical cloaking, and so on. In fact, apart from these uses, plasmonic materials can also be exploited as ENZ metamaterials when operating near their bulk plasma frequency, where the real part of the permittivity presents a near-zero value.

Therefore, ENZ metamaterials are achieved by using a certain plasmonic material at its plasma frequency.

There are many kinds of naturally occurring plasmonic materials, which play important roles in different applications. Noble metals, doped semiconductors, and some dielectrics are typical plasmonic materials with different plasma frequencies and also have significant differences in optical responses, enabling them to play fundamental roles in plasmonic materials and devices. Investigating the optical response of these plasmonic materials is crucial for facilitating their applications. It is known that the dielectric permittivity of the materials is related to the electric polarization, which represents the interaction between the electrons and the electromagnetic field. The real and imaginary parts of the dielectric permittivity $\varepsilon(\omega)$, denoted as ε_1 and ε_2 as a function of frequency, represent the polarization ability induced by the electric field and the losses from this process, respectively [62]. Generally speaking, lower loss (i.e., smaller imaginary part ε_2) is desirable for various applications of plasmonic materials. The optical response of these materials can be described by using the Drude model for free electron gases or the Lorentz model for bound electrons and vibrational resonances [15]. In fact, fundamental loss mechanisms of almost all plasmonic materials have a close relationship with conduction electrons and bound electrons (interband transition electrons). For conduction electrons, the losses usually arise from electron–electron and electron–phonon interactions, defects, surface states, and scattering from the lattice vibration. Since the conduction electrons have almost continuously available states, their behaviors when interacting with the electromagnetic field can be well predicted by using the classical Drude model, which regards conduction electrons as a free electron gas. These free electrons contribute to the negative real part of the permittivity of the plasmonic materials. The Drude model of plasmonic materials can be written as

$$\varepsilon_{r,\text{Drude}} = 1 + \chi - \frac{q^2 N}{\varepsilon_0 m} \frac{1}{\omega^2 + i\Gamma_i\omega}; \tag{3.1}$$

here, ε_0 is the free space permittivity, χ is the material susceptibility at the infinite frequency, Γ is the damping rate, N is the free carrier density, m is the effective mass of an electron, and q is the electron charge. The Drude model in Eq. (3.1) is an effective function for the permittivity of some plasmonic materials such as noble metals over a wide frequency range [63]. However, it cannot accurately predict the dielectric property of some noble metals such as gold and silver at some specific operating wavelengths. This difference is mainly attributed to the interband transition loss from bound electrons when excited by higher-energy photons, and this interband transition process is not taken into account in the Drude model. When considering the interband transition electrons, the dielectric permittivity of metals can be written as the superposition of many higher-order oscillation modes of the Lorentz model,

$$\varepsilon_{r,lorentz} = 1 + \chi + \sum_i \frac{A_i}{\omega_{o,i}^2 - \omega^2 - i\Gamma_i\omega} ; \tag{3.2}$$

here χ is the material susceptibility at the infinite frequency, A_i is the strength of the ith oscillator governed by the dipole transition rates, and $\omega_{o,i}$ is the resonant frequency. Fig. 3.1(a) and (b) show the Lorentz model and the Drude model, respectively, both illustrating the real and imaginary parts of the permittivity of the plasmonic materials. For the Lorentz model, the real part of the permittivity presents a positive value when the frequency is lower than the resonance point, with a negative valley slightly higher than the resonant frequency. It is also seen that an obvious absorption peak occurs at the resonant frequency. For both models, there is a frequency where the real permittivity goes across zero. The frequency regions where the real part of the media exhibits the epsilon-near-zero (ENZ) response are shaded in green ($|\text{Re}\{\varepsilon\}|$ < 1), and these occur around the resonant condition for each oscillator. In these regions, plasmonic materials can be exploited as ENZ metamaterials.

As seen from Fig. 3.1, even if the real part of the permittivity goes across the zero point, the imaginary part (i.e., material loss) still presents a nonzero value of permittivity, thus leading to significant influence on the ENZ performance. To tackle this issue, efforts have been made to find low-loss plasmonic

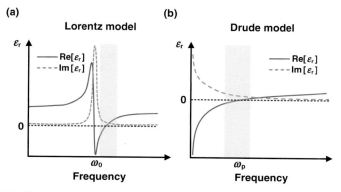

Figure 3.1 The real and imaginary parts of the permittivity described by (a) the Lorentz model and (b) the Drude model. The frequency regions where media exhibit the ENZ response are shaded in green.

materials. Actually, most of the plasmonic applications, especially with regard to ENZ metamaterials, need to overcome the losses of materials at the optical frequency regime. One possible way is to utilize the gain medium to compensate the large loss of the plasmonic materials. However, this method is not suitable for some subwavelength-scale applications due to high transparency current density [38, 39]. Another effective method is to reduce the free-electron density in metals or increase this value in semiconductors. Apart from the loss issue, the real part of the dielectric permittivity of plasmonic materials is another important factor needed when finding alternative low-loss plasmonic materials. For instance, an extremely negative permittivity of the plasmonic materials is not required for many subwavelength applications. Essentially, there are two key parameters that determine the optical response of plasmonic materials: the carrier concentration and carrier mobility (see Fig. 3.2) [64]. Enough carrier concentration ensures the suitable negative value of the dielectric permittivity. On the other hand, the higher carrier mobility contributes to the lower Drude damping frequency, thus the lower loss of the plasmonic materials. The ideal candidates for plasmonic materials should be located at the upper part of the plot. Without the illustration in this figure, the interband loss mentioned previously is highly undesirable for applications.

As illustrated in Fig. 3.2, in the ultraviolet (UV) wavelength range noble metals such as Au and Ag, and alkali metals such as K and Na are suitable plasmonic materials for ENZ applications. Transition-metal nitrides such as titanium nitride (TiN) can be good substitutes for traditional metals in the

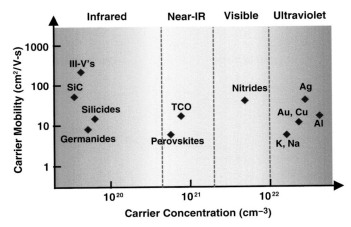

Figure 3.2 Material space of plasmonic materials. Important material parameters – carrier concentration and carrier mobility – form the optimization phase space. Data are used from [64].

visible frequency regime. Transparent conducting oxides (TCOs) such as indium-tin-oxide (ITO) are options for plasmonic applications in near-infrared (NIR) wavelengths due to their intermediate carrier concentration. In addition, silicides, germanides, and other semiconductors are suitable for the mid-infrared (MIR) wavelengths [64, 65]. Also, the ENZ effect or ENZ phenomenon can be observed within subwavelength-thin films or slabs. For example, the ENZ electromagnetic mode was introduced and confined in a thin layer of 22 nm, which consists of a GaAs quantum well between two AlGaAs barriers (approximately $\lambda_0/1{,}560$), to engineer the large absorption [66]. A relatively large change in reflectivity can be experimentally controlled with the applied voltage. Control of the ENZ mode opens an avenue for the new class of optoelectronic devices such as modulators, sources, and detectors. Besides that, the ENZ mode can also be supported by an ultrathin plasmonic nanolayer at the plasma frequency, where the dielectric permittivity of this film is near zero [67]. The ENZ mode was actually part of the long-range surface wave mode (SPP mode), which had an almost-flat dispersion relation and a large electric field in this film. This work enhanced our understanding of the thickness range of the plasmonic film capable of supporting this so-called ENZ mode, paving the way for many applications such as directional perfect absorption and ultrafast voltage-tunable devices. Furthermore, in the theoretical work of [68], the nonlinear and nonlocal optical properties from the free electron gas were investigated in ENZ metamaterial slabs composed of two-dimensional (2D) arrays of metallic, cylindrical nanoshells excited by transverse-magnetic (TM) polarized light. These ENZ nanoshells can be compensated with low damping frequency by embedding a dielectric core, which is a combination of a dielectric and an active gain medium. Using this design, many intriguing effects, such as impedance matching and electric field enhancement, were induced and exploited to boost second-harmonic generation (SHG) efficiency. Recently, the topological insulator has become an important topic in material science due to its intriguing properties. Interestingly, the semiconductor $Bi_{1.5}$ $Sb_{0.5}Te_{1.8}Se_{1.2}$ (BSTS), also known as a topological insulator, is also a good plasmonic material in the blue–ultraviolet and visible wavelength ranges [69]. The topological insulator BSTS achieved plasmonic resonances from 350 to 550 nm, attributed to the topologically protected surface conducting state and bulk charge carriers due to interband transitions. This new method of achieving plasmonic properties using topological insulator semiconductors also enables the realization of ENZ metamaterials.

As we have explained, the plasmonic materials are a feasible way to achieve ENZ properties in nature. ENZ metamaterials can be achieved only by depending on existing plasmonic materials at a certain plasma frequency. It is a challenge to achieve ENZ metamaterials at any desired frequency. For this purpose, with regard to metamaterials, an artificial way can be used to achieve ENZ without the limitation of natural plasmonic materials or intrinsic plasma frequency. As an important development of metamaterials, the structural dispersion of guided-wave structures, including parallel-plate waveguides and rectangular waveguides, can be explored to emulate plasmonic materials only with conventional positive-permittivity materials [70]. The idea of structural dispersion is using properly designed boundaries made of regular conductive metals to realize the plasmonic property with lower loss, including the low-loss ENZ metamaterials in any desired operating frequency. In the work of [71], a new methodology to reduce the loss in ENZ metamaterials was theoretically and numerically demonstrated by taking advantage of structural dispersion in waveguides. By changing the width of the waveguide, the effective material parameters can be manipulated, leading to smaller loss. Two typical examples, including plane waves propagating within an ENZ medium and SPP waves propagating along the interface of an effective negative-permittivity medium, were investigated with longer propagation distance, providing a way toward the low-loss plasmonic materials. The real and imaginary parts of the ENZ materials also were investigated to find the influence on the propagation characteristics of electromagnetic waves. The fundamental principle of causality has shown that electromagnetic waves propagating in ENZ materials with very low (asymptotically zero) loss have an asymptotically zero group velocity. It has been analytically and numerically demonstrated that realistic ENZ materials have very high reflection losses and propagation losses at the plasma frequency [72].

3.2 Periodic Structures

Unlike plasmonic materials near the plasma frequency, the ENZ metamaterials can also be realized by periodic structures. The first group of periodic structures uses subwavelength-scale resonators. Through elaborately designing artificial electromagnetic metamaterials composed of subwavelength-scale resonators, such as split-resonant rings (SRR) or periodically stacked metal–dielectric structures, an effective ENZ response can be engineered and obtained. As illustrated in Fig. 3.3(a), a microwave metasurface consisting of periodic planar complementary split-ring resonators (CSRR) printed on a circuit board was utilized to form the effective ENZ metamaterials [19]. The metal in microwave frequencies can be treated as a perfect electric conductor (PEC), and the metallic loss here is much smaller than that in optical frequencies, in which the metal belongs to the plasmonic materials.

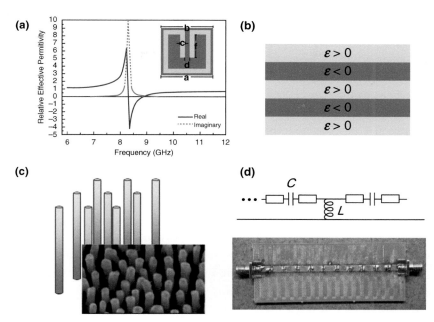

Figure 3.3 ENZ metamaterials constituted by periodic structures. (a) Microwave metasurface composed of complementary split-ring resonators [19]. (b) Alternatively stacked layers of epsilon-positive and epsilon-negative materials [75]. (c) Wire medium with effective Drude dispersion [23]. (d) Artificial transmission line loaded with series capacitance and shunt inductance [89]. Figures adapted with permission from: (a), [19], APS; (c), [23], APS; (d), [89], IEEE.

The second group of periodic structures uses plasmonic resonators to design ENZ metamaterials in the optical domain. A strategy for designing a broadband real part of near-zero effective permittivity was proposed with a flat multilayer metal–dielectric stacked structure [73]. The ENZ metamaterials usually operate at a single frequency or within a narrow frequency band, and it has become a challenge to design a broadband ENZ metamaterial. By using the Bergman–Milton representations discussed in [73], the broadband near-zero real part of the effective permittivity can be designed in spectral space in the form of a series of spectral factors, which can be translated into physical structures of the ENZ metamaterials by the classical mixing rule. In this research, multiple layers are utilized to generate multiple ENZ frequencies and link them together by stacking all the layers. In each layer, there is a periodic structure to form a single ENZ frequency. It can be designed by using an inverse algorithm to control the poles and zeros in the dispersion curve of the permittivity. For the same purpose, a wide-bandwidth, three-dimensional, zero-index metastructure has been theoretically and numerically demonstrated with low loss and low dispersion at visible frequencies [74]. Ag nanocube arrays were periodically embedded inside a dielectric material with identical lattice constant in three directions, and the low effective near-zero refractive index (< 0.2) achieved a relatively wide bandwidth of approximately 40 nm. An optical, three-dimensional metamaterial composed of a well-designed sculpted parallel array of subwavelength silver and silicon nitride nanolamellae was engineered to present a near-zero effective permittivity [75]. The ENZ response of this structured metamaterial can be tuned over a wide spectral range just by designing the metal filling ratio of the whole geometry (see Fig. 3.3(b)). This method has been exploited to achieve various optical devices such as hyperlenses for super-resolution imaging. In [76], the condition to implement an ENZ material, under which the electric displacement field vanished, was investigated in the presence of the spatial dispersion. The ENZ metamaterials made of periodic structures can exhibit strong nonlocality [23], [77, 78, 79]. As illustrated in Fig. 3.3(c), a wire medium composed of plasmonic nanorod anisotropic metamaterials was theoretically and experimentally analyzed in the optical regime to realize the ENZ response [23]. This wire composite medium was highly affected by the nonlocality, accompanied by excitation of the additional wave. The evidence of interference between the main and additional waves in this wire medium was given, and an analytical description of the optical response of the ENZ nanorod medium was presented.

Conventionally, metallic components or inclusions are usually employed to achieve the zero-index or negative-refractive-index metamaterials at the optical regime, leading to intolerable ohmic loss and thus hindering applications of the metamaterials. To tackle this severe problem, as the third group of periodic structures, low-loss dielectric resonators or photonic crystals are used to design ENZ metamaterials. By engineering the resonant mode of silicon rods, an impedance-matched zero-index metamaterial based on purely dielectric constituents was experimentally demonstrated at optical frequencies [80]. This designed all-dielectric, zero-index metamaterial was composed of stacked silicon-rod unit cells, which present nearly isotropic near-zero responses for a particular polarization, leading to angular-selectivity transmission and spontaneous emission. In this configuration, the band diagrams are usually utilized to predict the ENZ frequency with a certain mode. In the framework of photonic crystals, there are two approaches to realizing the near-zero-index effect, and their underlying physics and electromagnetic properties are different. The first approach is based on the Dirac-cone-shape dispersion (crossed linear dispersion) at the center of the Brillouin zone (Γ point) due to the accidental degeneracy [29], [81], where the photonic crystal behaves as an EMNZ medium, as its group velocity is finite and the effective intrinsic impedance in the medium is matched. The second approach is based on the band-edge effect [82, 83] of the photonic crystal with undegenerated electric or magnetic resonance. In this circumstance, the photonic band structure is flat at the Γ point, and the photonic crystal behaves as an epsilon-near-zero or a mu-near-zero medium, which exhibits zero group velocity and unmatched intrinsic impedance. The zero-refractive-index metamaterial or ENZ metamaterial can also be implemented by exploring the band structure of a two-dimensional dielectric photonic crystal. When the Dirac cone dispersion is at the center of the Brillouin zone, 2D photonic crystals consisting of a square dielectric-rod lattice can behave as if they had zero refractive index. This method can offer an effective zero permittivity and permeability with a relatively low loss compared with those using metallic components at high frequencies [29]. As discussed in [84], a 2D accidental-degeneracy photonic crystal consisting of a square lattice array of elliptical dielectric cylinders presented a semi-Dirac point located at the center of the Brillouin zone and an electromagnetic topological transition (ETT) phenomenon. An effective medium was exploited to explain this phenomenon and demonstrate that this photonic crystal possessed either a zero-refractive index or an epsilon-near-zero property at the semi-Dirac point. To achieve photonic crystals with zero index as a desired topological property, a large-scale computational technique, also known as topology optimization, was developed and applied to the inverse design of the photonic Dirac cone [85]. By this method, various photonic crystal geometry structures exhibiting dual-polarization cones can be realized using only simple and isotropic dielectric materials. The optimization technique developed in this research provided a new way to engineer band structure and control electromagnetic waves in periodic materials. Based on this method, on-chip integrated zero-index metamaterials were also investigated in [30]. An on-chip integrated zero-index metamaterial in optical frequency was designed and fabricated using a square array of low-aspect-ratio silicon pillars on a silicon-on-insulator substrate. The pitch and radius of the silicon pillar array can be varied to construct a Dirac-like cone at the Brillouin zone of the band structure to achieve near-zero index with low loss. This method was the first attempt to exploit near-zero index metamaterials in integrated optics.

Although the photonic crystals whose Dirac cone is at the center of the Brillouin zone realize the zero-index property, this methodology usually produces strong and undesirable radiation into the free space due to the inherent symmetry, resulting in large losses in the realization of zero-index metamaterials. To solve this problem, a photonic crystal slab with zero-index modes that can eliminate the radiation loss was proposed, which is attributed to the symmetry-protected bound states in continuum [86], and we categorize this as the fourth group of periodic structures for ENZ metamaterials. Associated out-of-plane radiation loss was completely eliminated for a zero-index metastruc-ture, obtaining a low propagation loss of 1 dB/mm and an effective index below 0.1 within a 10 nm bandwidth in the telecommunication band around wave-length λ = 1,550 nm. Radiative loss in near-infrared, zero-index, on-chip photonic crystals was also mitigated by increasing the total quality factor by an order of magnitude [87]. By introducing resonance-trapped and symmetry-protected states, quality factors of 2.6×10^3 and 7.8×10^3 respectively were experimentally obtained at near-infrared wavelengths. Moreover, the zero-index photonic crystal with a 10-fold loss reduction can achieve on-chip integration, impedance matching, and scalability over the visible and infrared frequencies, contributing to the low loss and long propagation length in pho-tonic integrated circuits. In [88], an ultralow-loss, on-chip integrated Dirac cone material, composed of a square array of silicon pillars embedded in silicon dioxide, was achieved by constructing destructive interference around the material structure. The designed zero-index photonic crystal slab in this work presented a propagation loss as low as 0.15 dB/mm at the zero-index frequency, where the refractive index was near zero ($|n_{\mathrm{eff}}| < 0.1$) over a bandwidth of 4.9%.

Last but not least, as the fifth group of periodic structures for ENZ metama-terials, an artificial electromagnetic transmission line formed by periodically cascaded metamaterial unit cells also was studied [89, 90, 91], and it has potential applications in microwave devices. With series capacitors and shunt inductors, the ENZ transmission line was experimentally demonstrated in Fig. 3.3(d), which can also be applied to zero-degree, phase-shifting applica-tions [89]. This method enables the planar integrated physical realization of the ENZ metamaterials.

3.3 Waveguide ENZ Metamaterials

We have introduced two common methods of realizing ENZ metamaterials by using natural plasmonic materials and artificial periodic structures, respectively. Actually, apart from these two methods, ENZ metamaterials can also be phys-ically realized based on the structural dispersions of waveguides. Next, we introduce the concept and design method of waveguide ENZ metamaterials, which have low-loss and non-plasmonic merits. Approximately four decades ago, Rotman found and proved that parallel-plate waveguides carrying the fundamental TE_{10} mode can emulate plasmonic materials at the microwave frequency regime, when the waveguide is operating below its cutoff frequency of TE_{10} mode [26]. According to this property, a parallel-plate waveguide medium below cutoff was used as an epsilon-negative (ENG) metamaterial to reduce the total scattering cross section of a two-dimensional dielectric in a specific frequency band [92]. For a TE_{10}-mode parallel plate waveguide with a width of h and filled dielectric of permittivity ε_r, electromagnetic wave propagation on the middle cut plane of the waveguide is illustrated in Fig. 3.4(a). The wave propagation along the middle plane inside the waveguide can be made an analogy to plane-wave propagation in an equivalent homogenous material with effective permittivity of $\varepsilon_{eff} = \varepsilon_r - c^2/4h^2f^2$, which presents a dispersion related to the waveguide structure parameters, as shown in Fig. 3.4(b). In this way, we use a 2D uniform plane wave (i.e., transverse electromagnetic (TEM) wave) on the middle plane to imitate the TE_{10}-mode inside the waveguide. Indeed, the virtual 2D TEM wave is part of the actual 3D TE_{10}-mode wave, with an identical propagation constant and longitudinal phase progress. In addition, the propagation constant of electromagnetic waves within this waveguide is written as $\beta = ((2\pi nf)^2 - (\pi/h)^2)^{1/2}$ [93] . When operating at the cutoff frequency of its TE_{10} mode, the waveguide can emulate the metamaterial with the effective dielectric permittivity $\varepsilon_{eff} \approx 0$ and propagation constant $\beta \approx 0$. Under these conditions, the electromagnetic wave within the waveguide demonstrates an infinite phase velocity and static-like field distribution. In addition, a rectangular waveguide (as shown in Fig. 3.4(c)) and its planar-form substrate-integrated waveguide (SIW, as shown in Fig. 3.4(d)) [94] also can

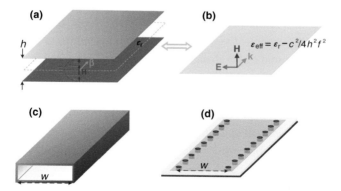

Figure 3.4 Waveguide-emulated ENZ metamaterials. The wave propagation on the middle cut plane of a parallel-plate waveguide under the cutoff TE_{10} mode (a) is equivalent to the plane wave propagation in the homogenous medium with $\varepsilon_{eff} \approx 0$ [26]. (b). Rectangular waveguide (c) and the substrate-integrated waveguide (d) can also emulate the ENZ response at their cutoff frequency of the TE_{10} mode.

mimic ENZ metamaterial when operating at their own cutoff frequency of TE$_{10}$ mode.

This method of achieving effective ENZ metamaterials can also be realized in the optical regime. A light-guiding nanostructure using a metal-insulator-metal structured waveguide at cutoff was fabricated and experimentally characterized to verify the effective permittivity $\varepsilon = 0$ and effective refractive index $n = 0$ [95]. This plasmonic nanowaveguide may pave the way for building integrated circuits in nanoscale. Silicon-based zero-index waveguides supporting phase-free propagation in telecom frequencies were proposed and experimentally verified in [96]. Such a waveguide would consist of a row of zero-index metamaterial corrugated with periodic holes on both its sides, surrounded by a photonic band gap material. This design can easily be integrated with silicon photonic circuits or chips. As a consequence of waveguide ENZ metamaterials, we can use a regular dielectric inside a waveguide to realize all the ENZ properties at any frequency with lower loss than plasmonic materials. It should be noted that, for microwave frequencies, we have metal as the PEC boundary. When the frequency goes up to terahertz, the metal is no longer a PEC, with waves penetrating inside the metal. For frequencies higher than 150 THz [37], the metal is no longer good as a PEC boundary and behaves as a dielectric with negative permittivity. To solve this problem, a periodic photonic crystal structure can be used to make the boundaries for structural dispersion [42].

It is known that plasmonic materials, including noble metals such as silver and gold, semiconductors, and dielectrics, have been exploited to achieve a negative permittivity at optical frequencies due to their inherent Drude-like or Lorentz-like dispersion response. However, as mentioned earlier, many plasmonic materials exhibit serious losses due to the conduction electrons and bound electrons (interband transition electrons). Therefore, low-loss plasmonic materials have been continuously pursued and investigated, including highly doped semiconductors, transparent conducting oxides (TCO), gain-assisted materials, and so on. Despite these efforts, loss is still the most important issue for plasmonic-related technologies and applications. In [71], a different methodology to reduce the loss in plasmonic materials was introduced by exploring the structural dispersion in waveguides. As illustrated in Fig. 3.5(a), the structural dispersion-based reduction of loss was achieved by taking advantage of a parallel-plate waveguide (PPW) with a width of d. For simplicity, a waveguide made of perfect electric conductor (PEC) walls (denoted in yellow) was filled with a lossy plasmonic material, whose relative

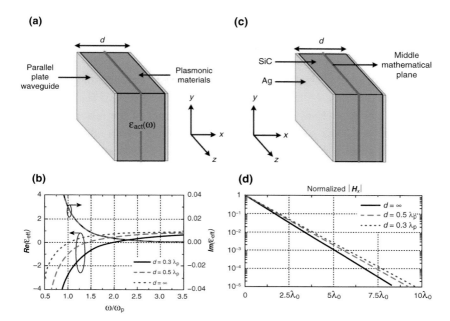

Figure 3.5 Generic representation of structural-based loss mitigation method [71]. (a) Lossy plasmonic materials sandwiched by a parallel-plate waveguide (PPW); the solid red line presents the middle mathematical plane. (b) The real and imaginary parts of the relative effective permittivity of the plasmonic materials within a PPW of various widths d. Here, the frequency is normalized by the plasma frequency ω_p. (c) The ENZ medium, that is, SiC material using the Lorentzian model bounded by an Ag PPW using the Drude model. (d) Normalized magnitude of Hx along the z direction. (a)–(d) are adapted with permission from [71], AAAS.

permittivity was described by the Drude model (here, $\exp(-i\omega t)$ time convention is assumed),

$$\varepsilon_{\text{act}}(\omega) = 1 - \omega_p^2/\omega(\omega + i\gamma), \tag{3.3}$$

where ω_p is the plasma angular frequency, ω is the operating frequency, and γ is the collision frequency. The real and imaginary parts of the ε_{act} are represented in Fig. 3.5(b) by the blue dotted curves. For the PPW shown in Fig. 3.5(a), the propagation constant k of the TE_{10} mode can be expressed as

$$k = \sqrt{k_0^2\varepsilon_{\text{act}}(\omega) - (\pi/d)^2} = \omega\sqrt{\varepsilon_{\text{eff}}(\omega)\varepsilon_0\mu_0}, \tag{3.4}$$

and the relative effective permittivity can be written as

$$\varepsilon_{\text{eff}}(\omega) = \varepsilon_{\text{act}}(\omega) - c^2\pi^2/\omega^2 d^2, \tag{3.5}$$

where c is the speed of light in free space. The relative effective permittivity values for width of $d = 0.3\lambda_p$, $d = 0.5\lambda_p$ ($\lambda_p = 2\pi c/\omega_p$) are illustrated in Fig. 3.5(b), calculated from Eq. 3.5. It is clearly seen that only the real part of the relative effective permittivity is affected when varying the width of the PPW. In other words, the real and imaginary parts of the relative permittivity can be independently controlled, thus providing a new way to reduce the loss of plasmonic materials. Two examples were demonstrated in [71] to verify the proposed methodology, including the wave propagation in epsilon-near-zero media and the propagation of surface plasmon polaritons. Here, only the wave propagation in ENZ metamaterials (i.e., plasmonic materials with a near-zero permittivity) are presented in Figs. 3.5(c) and (d). The complex wave number $k(\omega, d, \gamma)$ in the lossy plasmonic materials can be written as

$$k(\omega, d, \gamma) = \omega\sqrt{1 - \omega_p^2/\omega(\omega + i\gamma) - c^2\pi^2/\omega^2 d^2}/c = \beta + i\alpha. \tag{3.6}$$

For $\omega = \omega_{\text{ENZ}}$ (i.e, the real part of the relative effective permittivity is zero), the complex wave number can be obtained as

$$k(\omega, d, \gamma) = \omega_{\text{ENZ}}\sqrt{\text{Im}[\varepsilon_{\text{eff}}(\omega_{\text{ENZ}}, d, \gamma)]}\,(1 + i)/c\sqrt{2}. \tag{3.7}$$

At the ENZ frequency ($\omega = \omega_{\text{ENZ}}$), the phase ($\beta$) and attenuation ($\alpha$) constants are given by

$$\beta = \alpha = \omega_{\text{ENZ}}\sqrt{\text{Im}[\varepsilon_{\text{eff}}(\omega_{\text{ENZ}}, d, \gamma)]}\,/c\sqrt{2}. \tag{3.8}$$

Fig. 3.5(c) presents the three-dimensional PPW structure filled with ENZ medium, which is realized by the plasmonic material SiC. Here, the material of PPW is Ag, whose relative permittivity can be calculated by the Drude model [71]. The relative permittivity of the SiC can be calculated using the Lorentzian model. Therefore, the ENZ frequency of SiC occurs at 29.13 THz, where the real part of the effective permittivity is zero, while the imaginary part is 0.101. The spacing between the two plates is d, and the thickness of the two Ag plates is $\lambda_0/5$. The overall length of the structure along the z-axis is $10\lambda_0$. Two different widths of $d = 0.5\lambda_0$ and $d = 0.3\lambda_0$ were studied, with the theoretical ENZ frequency calculated at 29.93 and 31.65 THz, respectively, where the Ag material can still be a good conductor. As a result, $\beta = \alpha = 0.201k_0$ at 29.93 THz for $d = 0.5\lambda_0$, and $\beta = \alpha = 0.165k_0$ at 31.65 THz for $d = 0.3\lambda_0$ can be calculated from Eq. 3.8. From Fig. 3.5(d), it is seen that the field intensity of normalized magnitude of H_x along the z axis becomes stronger with the

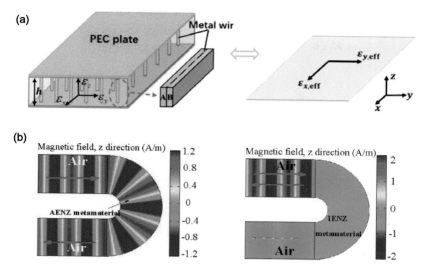

Figure 3.6 Anisotropic ENZ metamaterials in waveguides. (a) Implementation of the anisotropic ENZ metamaterial via a waveguide filled with layered slabs (left side) and its equivalent homogenous medium (right side). (b) The magnetic field configurations over the bent waveguide channels filled with anisotropic (left side) and isotropic (right side) ENZ metamaterials. (a) is adapted from [101], OSA, and (b) is adapted from [99], AIP.

narrower width d of the PPW due to the reduced attenuation constant. The full-wave numerical results agree well with the theoretical prediction, verifying the effectiveness of the proposed methodology in [71].

The aforementioned waveguide structures are mainly related to the effective realization of isotropic ENZ metamaterials on a specific plane (e.g., the horizontal plane of a parallel-plate waveguide). On the other hand, the anisotropic ENZ (MNZ) metamaterials [12], [97]–[102] with one or several elements in the permittivity (permeability) tensor close to zero have also been investigated in scenarios of waveguides and free space. They have been demonstrated to span a wide range of functionalities, including arbitrary control of energy flux [97], perfect bending waveguides [98]–[101], scattering suppression [102], and so on. As an exemplary design of the waveguide-based anisotropic ENZ metamaterials depicted in Fig. 3.6(a), the seminal work [101] proposed to employ the waveguide filled with layered media to realize the anisotropic permittivity in the horizontal (x-y) plane. The anomalous tunnelling effect for the channel filled with anisotropic ENZ medium is quite different from the case of isotropic ENZ medium. As demonstrated in Fig. 3.6(b), the bent waveguide channel filled with anisotropic ENZ medium hardly incurs any reflection of waves [99], even though the channel is not extremely narrow. Detailed theoretical analysis can be performed from the perspective of transformation optics [100].

3.4 Photonic Doping

In this section, we introduce the last design method for ENZ metamaterials, as well as one of the most intriguing applications of the ENZ properties. Based on the ENZ property realization in Sections 3.1 to 3.3, we can further tune the effective permeability of the ENZ medium operating as the ENZ-based metamaterials. To supply some background, it is known that doping is a concept stemming from semiconductor and integrated circuits process areas, where it enables the electrical, optical, or magnetic properties of the materials by introducing some foreign impurities. For example, by doping pentad phosphorus into quadrivalent silicon, the free electrons of phosphorus lead to the transition of silicon from an insulator to a conductor. The doped phosphorus is random but changes the property of silicon as an N-type semiconductor. Interestingly, the concept of doping can be effectively transplanted to the fields of macroscopic photonics [31], also known as photonic doping, which can realize matched zero-index metamaterial [41] (i.e., EMNZ metamaterial), or total transmission with defects [103]. Before the concept of photonic doping formally came into shape, several preceding experiments [104]–[109] revealed the control of reflection, transmission, and scattering by using defects in the zero-index media. For example, Y. Lai's group demonstrated the total reflection effect for the ENZ body containing a dielectric defect and behaving as a perfect magnetic conductor [104], and even proposed the solution to obtain Fano resonances in the ENZ metamaterials with embedded layered impurities [109]. The work by S. L. He's group investigated how to enhance or suppress the radiation of a source in MNZ media via including PEC structures [105].

The detailed theory of 2D photonic doping is included in [31], and here is a brief review of this work. As illustrated in Fig. 3.7(a) and (b), when immersed in a two-dimensional, arbitrarily shaped ENZ metamaterial, macroscopic dielectric particles behave as dopants that can influence the effective permeability of the whole ENZ medium without affecting its effective permittivity. Moreover, this phenomenon is independent of dopants' position within the ENZ host medium and the shape of the ENZ host medium. It is noted that the incident electromagnetic wave is polarized with the magnetic field along the out-of-the-plane axis. As seen from the outside world, the 2D ENZ metamaterial behaves as a homogenous material with the effective permeability given by [31]:

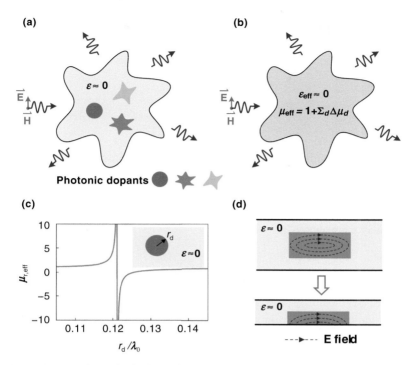

Figure 3.7 2D photonic doping of ENZ metamaterials. (a) Concept of photonic doping [31]: a 2D arbitrarily shaped ENZ body comprising 2D macroscopic dielectric impurities, exposed to the incidence polarized with the magnetic field along the out-of-the-plane axis. (b) The equivalent homogeneous 2D ENZ body with a uniform effective permeability μ_{eff} as seen from the outside world. (c) Calculated effective permeability of a 2D ENZ body with an area of $0.5\lambda_0{}^2$, which comprises a dielectric circular rod with a relative permittivity of 10 and a radius of r_d. (d) Principle of the substrate-integrated photonic doping from the perspective of electric-field distribution analysis [32].

$$\mu_{\text{eff}} = \left(A - \sum_d A_d + \sum_d \mu_d \iint_{A_d} \psi^d \mathrm{d}s \right) / A, \tag{3.9}$$

where A and A_d represent, respectively, the total cross-sectional areas of the ENZ metamaterial and the cross-sectional areas of photonic dopants; μ_d denotes the relative permeability of the dopants; and $\psi^d(r)$ is the normalized magnetic field subject to the wave equation and boundary condition:

$$\nabla^2 \psi + k_d^2 \psi^d = 0, \quad \psi|_{\partial A_d} = 1, \tag{3.10}$$

where k_d is the wave number in the dopant. We present here a brief derivation of the effective permeability formula, Eq. (3.9), from the basic point of view of Maxwell's equations. Apply Faraday's law on the boundary of the ENZ meta-material to yield

$$\oint_{\partial A} \mathbf{E} \cdot \mathbf{dl} = -j\omega \iint_A \mathbf{B} \cdot \mathbf{ds} = -j\omega\mu_0 H_0 (A - \sum_d A_d + \mu_d \sum_d \iint_{A_d} \psi^d ds),$$

(3.11)

where ω is the angular frequency; $\mathbf{E}$ and $\mathbf{B}$ denote respectively the electric field and magnetic flux density; and H_0 stands for the magnetic field in the ENZ host, which is position independent. If the doped ENZ metamaterial were replaced by a homogenous ENZ medium with a relative permeability of μ_{eff}, then the result of Faraday's law would be

$$\oint_{\partial A} \mathbf{E} \cdot \mathbf{dl} = -j\omega \iint_A \mathbf{B} \cdot \mathbf{ds} = -j\omega\mu_0\mu_{\text{eff}}H_0.$$

(3.12)

Comparing Eqs. (3.11) and (3.12), we can easily obtain the expression of effective permeability μ_{eff}, Eq. (3.9), of the ENZ metamaterial comprising photonic dopants. Although photonic dopants in principle are allowed to take arbitrary shapes, dopants with high geometrical symmetry are much more convenient for theoretical analysis. Consider a circular photonic dopant with a radius of r_d embedded in the ENZ host; the normalized magnetic field in the dopant has been derived analytically as [31]:

$$\psi(\mathbf{r}) = J_0(k_d r)/J_0(k_d r_d).$$

(3.13)

Due to the homogenous background magnetic field in the ENZ host, only the electromagnetic mode described by axial-symmetric cylindrical harmonics is allowed. By substituting Eq. (3.13) into Eq. (3.9), the formula of effective permeability is therefore derived explicitly as

$$\mu_{\text{eff}} = \frac{1}{A}\left[\frac{2\pi r_d}{k_d}\frac{J_1(k_d r_d)}{J_0(k_d r_d)} - \pi r_d^2\right] + 1.$$

(3.14)

Fig. 3.7(c) plots the theoretical effective permeability of a 2D ENZ body with the area $A = 0.5\lambda_0^2$ (λ_0 is free-space wavelength), which comprises a circular dielectric rod ($\varepsilon_d = 10$, $\mu_d = 1$), as a function of radius r_d [31]. It is evident that the effective permeability of the 2D ENZ metamaterial can be arbitrarily designed by selecting a suitable radius of the doped dielectric rod. Moreover, an obvious resonance occurs where the effective permeability of the whole 2D ENZ medium μ_{eff} reaches infinity. At this time, the host body demonstrates itself as a PMC boundary. The equivalence of a doped ENZ medium and a perfect magnetic conductor body has been theoretically demonstrated in [104]. Another case allowing for an analytical solution is the ENZ host comprising a rectangular dopant. In [32], the Green's function of the Helmholtz equation in the rectangular dopant region was derived using a Fourier basis expansion,

$$G(x,y \; ;x',y') = \sum_{m,n=1}^{+\infty} \frac{U_{m,n}(x',y')U_{m,n}(x,y)}{\lambda_{m,n}{}^2 - k_d{}^2}, \tag{3.15}$$

where the basis function $U_{m,n}(x,y)$ and corresponding eigenvalue $\lambda_{m,n}$ are given as

$$\begin{aligned} U_{m,n}(x,y) &= \sqrt{4/(h_d l_d)} \cos(m\pi x/l_d)\cos(n\pi y/h_d) \\ \lambda_{m,n} &= \sqrt{(m\pi/l_d)^2 + (n\pi/h_d)^2} \end{aligned}, \tag{3.16}$$

where l_d and h_d denote the side lengths of the rectangular dopant. Once the Green's function is solved, the normalized magnetic field distribution can readily be derived as

$$\psi(x,y) = 1 + \sum_{m=1,n=1}^{+\infty} k_d{}^2 \frac{4((-1)^m - 1)((-1)^n - 1)}{\pi^2 mn} \frac{\cos(m\pi x/l_d)\cos(n\pi y/h_d)}{(m\pi/l_d)^2 + (n\pi/h_d)^2 - k_d{}^2}, \tag{3.17}$$

and the effective permeability for the rectangular dopant case is then obtained by inserting Eq. (3.17) into (3.9), which gives

$$\mu_{\text{eff}} = 1 + \sum_{m=1,n=1}^{+\infty} \frac{2l_d h_d((-1)^m - 1)^2((-1)^n - 1)^2}{A\pi^4 m^2 n^2} \frac{k_d{}^2}{(m\pi/l_d)^2 + (n\pi/h_d)^2 - k_d{}^2}. \tag{3.18}$$

As clearly shown in Eq. (3.18), the effective permeability diverges at the resonance frequencies of the TM$_{m,n}$ modes of a rectangular cavity, where the doped ENZ metamaterial behaves as a perfect magnetic conductor body. Existing schemes for proof-of-concept verification of the ENZ host and photonic doping have a bulky geometry or a high fabrication cost, and so they are difficult to apply in low-cost and space-limited systems. In Fig. 3.7(d), substrate-integrated photonic doping, implemented by using a printed-circuit-board integrated design, was proposed, with the height of the whole ENZ structure significantly reduced by half, a reduction attributed to the symmetric principle [32].

To get a better sense of 2D photonic doping for the magnetic field manipulation, we present a design example of an irregularly shaped ENZ

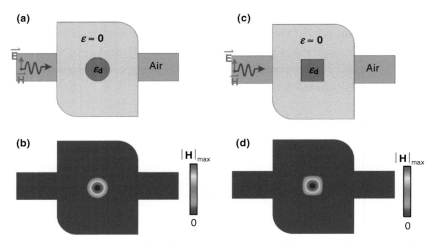

Figure 3.8 Design example of 2D photonic doping [31]. (a) Configuration of an ENZ medium comprising a circular dielectric dopant with a radius of 3 mm and relative permittivity of 40, and (b) simulated magnitude of the magnetic field distribution at the EMNZ tunneling frequency. (c) Configuration of an ENZ medium comprising a square dielectric dopant with a side length of 5.6 mm and relative permittivity of 40, and (d) simulated magnitude of the magnetic field distribution at the EMNZ tunneling frequency. The cross-sectional area of the ENZ media in both cases is 470 mm^2.

medium comprising a cylindrical dopant in Fig. 3.8(a), and the simulated magnetic amplitude distribution at the EMNZ tunneling frequency in Fig. 3.8(b). The cases for a square dopant and the corresponding magnetic field distribution are presented in Fig. 3.8(c) and 3.8(d), respectively. As seen, the magnetic field is highly enhanced in the dopant region, which indicates that the magnetic energy of the doped system is mainly stored in the photonic dopant. In the framework of photonic doping, one arbitrarily located photonic dopant is able to impact the effective constitutive parameter of the whole ENZ metamaterial. Such a paradigm essentially differs from that of periodical metamaterials, where it is an array of interacting meta-elements that contributes to the macroscopic response of the composite medium.

Electromagnetic transmission through a 3D EMNZ body embedded with random impurities also was investigated [110]. It was first proved and numerically demonstrated that any single impurity with finite volume embedded inside a 3D EMNZ host cannot influence the total transmission [110]. Furthermore, an unusual electromagnetic percolation behavior of photons squeezing through the gaps between random inclusions with unity transmittance was observed in this research. Importantly, such a percolation effect features a threshold induced by the long-range connectivity of the nonconducting component in the transverse direction and related to the geometry concept of "free surface."

In [43], by exploiting ENZ metamaterials or zero-index metamaterials, geometry-invariant resonant cavities whose eigenfrequencies were independent of their geometrical boundary deformations were theoretically investigated. For the applications of photonic doping, in most studies on photonic doping, non-absorptive or low-loss particles are usually adopted to tune the effective permeability of the ENZ host medium. On the other hand, absorptive defects can also be used to dope the epsilon- and mu-near-zero (EMNZ) metamaterials to achieve coherent perfect absorption, which was independent of the size and shape of EMNZ metamaterial as well as the position of the doping defects [111]. In contrast, the electromagnetic properties of bulk ENZ metamaterials can also be immune to the photonic doping effect of the impurities by exploiting a slab of ENZ metamaterial sandwiched with a pair of parity-time (PT)-symmetric metasurfaces. When the permittivity of the slab converges toward zero, only one exception point exists for the PT-symmetric system. As a result, extraordinary properties such as perfect transmission can be implemented independently of the material and shape of the embedded impurities [112]. A reconfigurable ENZ metasurface was also numerically proposed and investigated by taking advantage of the intriguing tailoring capability of the doped particles in ENZ metamaterials [113]. By tuning the properties of a few arbitrarily immersed dielectric particles, tunable photonic doping can be used to reconfigure the macroscopic electromagnetic properties of the ENZ host medium. In most applications of photonic doping, lossless or low-loss impurities are usually employed for intriguing electromagnetic applications such as perfect transmission. Actually, photonic doping can also exploit non-Hermitian configurations characterized by gain-loss distribution for the dopants or the host medium [114]. For example, when nonmagnetic core (gain)-shell (loss) or core (loss)-shell (gain) cylindrical particles are doped, the ENZ host medium can obtain a large tunability in the sign and amplitude of the effective permeability, opening an avenue for novel reconfigurable nanophotonic devices. In addition, when the ENZ host medium has a PT-symmetric gain-loss bilayer configuration, wave-guiding phenomena can be enhanced and tailored at the gain-loss interface. Apart from being applied to electromagnetic ENZ metamaterials, the concept of doping can also be transplanted to an acoustic area. For example, acoustic doping was theoretically, numerically, and experimentally investigated for the plate-type density-near-zero (DNZ) acoustic metamaterial structure [115]. By doping an elaborately designed Helmholtz resonator element into the DNZ metamaterial, a geometry-independent density-and-compressibility-near-zero (DCNZ) medium was achieved with the phenomenon of tunneling effect.

In this section, we have reviewed various realization methods of ENZ metamaterials, or ENZ-based metamaterials. A comprehensive comparison among different realization methods is addressed here. The plasmonic media are a kind of naturally occurring index-near-zero media, which is due to the collective resonance of electrons coupled to the excitation of electromagnetic waves at specific frequencies. The latter three methods, including waveguide, periodic structures, and photonic doping, on the other hand, resort to artificial structures emulating index-near-zero response at custom-designed frequency regions. Among those schemes, the waveguide-emulated ENZ metamaterials employ the analogy between the cutoff mode and the spatially static field configuration to mimic the ENZ response, while periodic metallic elements or photonic crystals harness the resonance of the coupled elements or special band structures to attain index-near-zero effects. The scheme of photonic doping opens up a pathway to accurately controlling the effective permeability and macroscopic response of the ENZ host via arbitrarily located foreign impurities, which can represent a nonperiodic paradigm of artificially structured media.

In Section 4, the central part of this Element, we aim to demonstrate the unusual applications of ENZ metamaterials to control electromagnetic waves and manipulate wave–matter interactions. Concretely, our discussion will focus on applications in microwave engineering, optics, and quantum physics regimes. Importantly, the physics underlying these intriguing metamaterial applications will be analyzed.

4 Applications of ENZ Metamaterials in Microwave and Terahertz Engineering

As discussed in Sections 2 and 3, the ENZ metamaterial supports an unusual zeroth-order resonance, which has two critical properties: geometry-independent frequency and uniform field distribution in both magnitude and phase for the 2D scenario. These interesting characteristics have been realized by using waveguides, which are important transmission lines in microwave engineering. From the other side, the ENZ metamaterial also enables new methods for microwave engineering, such as antennas, absorbers, and circuits. In this section, we focus on the microwave applications of ENZ metamaterials.

4.1 ENZ Metamaterial Antennas

Antennas are indispensable elements in wireless systems for the usage of transmitting and receiving electromagnetic wave signals. Antennas take energy from RF circuits and radiate it to free space. Therefore, two classes of parameters related to antennas are widely studied. The first one is the circuit-related parameters, such as input impedance, reflection, and transmission coefficients between circuit and antenna port. The second one is the radiation-related parameters, such as radiation patterns, directivity, gain, and so on. As widely discussed in antenna textbooks, the dimensions of the antenna determine the operating frequency and radiation patterns. For example, for a microstrip antenna, the operating frequency is calculated by the length of the antenna, usually resonant at the length of half wavelength. By changing the length, the radiation patterns also change with different distances between the two equivalent magnetic currents of a microstrip antenna. In this section, we discuss new antenna designs based on ENZ metamaterials from three aspects: (1) antenna miniaturization with electrically small dimensions; (2) antenna radiation pattern shaping and harmonic frequency tuning but with the ENZ frequency unchanged, also known as geometry-irrelevant antennas; (3) uniform field distributions for high-gain antennas, where the ENZ metamaterials perform not only as main radiators but also as directors and reflectors. For this direction, new materials or new structures with ENZ properties may lead to new opportunities for antenna designs.

Unlike ordinary antennas operating at a certain frequency determined by the dimensions, a new type of miniaturized antenna can be designed with the assistance of ENZ metamaterials, which are also known as zero-order resonant (ZOR) mode. For ZOR antennas, the resonant field inside the antenna is uniform. The resonant frequency of such an antenna can be independent of the dimensions, which do not affect the uniform field distribution, providing new freedom to design electric small antennas. In this way, the circuit-related parameters at the resonant frequency change only a little, but the quality factor and radiation patterns change according to the decreased physical dimensions of antennas. A low-profile ZOR antenna is proposed in [116] by loading serial interdigital capacitors and shunt meander-line inductors. The proposed antenna is miniaturized to 25% of the ordinary patch antenna when comparing the footprints. By replacing the distributed capacitors with varactors, a frequency-reconfigurable antenna based on ENZ resonance is composed in [117] to overcome the narrowband nature of this kind of antenna. Besides the CRLH structure, [118] proposes an electrically small MNZ monopole using rings loaded by interdigital slits with a 54.46% size reduction. To deal with the high loss introduced by substrates, a high-efficiency solution can be found in [119] where MNG metamaterials are realized by helical metal wires. By combining this MNG material with air, an MNZ metamaterial is composed and used for the miniaturization of patch antennas, which is only 40% of the normal one. The proposed antenna has a fully metallic structure so that it has high radiation efficiency. Besides, this miniaturization method also applies to optical bands where subwavelength particles are used as nanoantennas under the plasmonic resonance [120] to shape the radiation patterns in nanoscale.

In addition to antenna miniaturization, this geometry-irrelevant resonance may inspire more interesting properties in antenna design. Not only can we reduce the antenna's dimensions, we can also increase them for other purposes, such as harmonic wave frequency tuning and radiation pattern shaping. For harmonic wave frequency tuning, as we know, designing multiband antennas usually suffers from the difficulty of tuning each band independently. With the help of ENZ metamaterial, [121] proposes a dual-band patch antenna with flexibly controlled frequency ratio by partially loading MNZ metamaterial. Moreover, since the ZOR mode is always kept in whatever geometries, the height-to-width ratio can be manipulated for optimal impedance matching. For the radiation pattern shaping, as shown in Fig. 4.1(a), [122] proposes an omnidirectional antenna by using an ultra-thin ENZ slab to match the coaxial probe to the thick waveguide, which can efficiently radiate power into free space omnidirectionally for each angle. In this way, other types of radiation patterns can also be achieved by properly designing the radiating apertures. Another application of the ENZ metamaterial is to design antennas with tunable lengths at a fixed frequency. In Fig. 4.1(b), mushroom-shaped CRLH units are used to implement a ZOR antenna with a monopolar radiation pattern [123]. With a variable length from 0.233 λ_0 to 0.583 λ_0, the operating frequency remains near 10 GHz without obvious changes. Figs. 4.1(c) and (d) depict ENZ antennas based on the cutoff mode of the substrate integrated waveguide (SIW). In Fig. 4.1(c), an open-ended SIW is fed by a coaxial probe to obtain a dual beam with tunable directions and gains [124]. To achieve a broadside radiation, which is more widely used in communication systems, a ceramic block of half-wavelength is used to generate two in-phase magnetic currents at the opening end of the SIW, inspired by the concept of photonic doping in [125], as exhibited in Fig. 4.1(d). This length-independent property is useful in wearable and biometric applications where folding and stretching of the antenna are inevitably demanded. To realize a flexible implementation of this antenna, [126] proposed a fabrication technique based on a flexible substrate called PDMS and launched on-body tests with a stretchable ENZ antenna, demonstrating the usefulness and effectiveness of this unique property. In [118–126], even though the resonant frequency doesn't change with dimension, the radiation pattern at this frequency changes. In other words, the radiation patterns can be properly synthesized in this way without changing the resonant frequency.

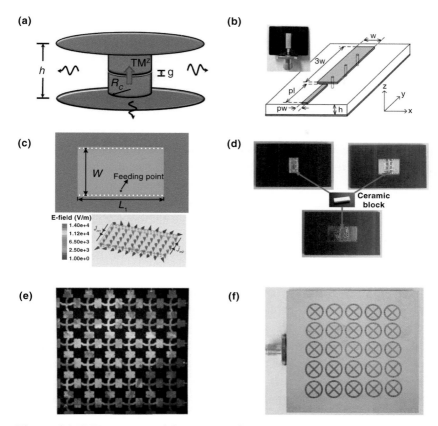

Figure 4.1 ENZ-metamaterial-empowered antennas. (a) Circular disk antenna matched by an ENZ metamaterial channel [122]. (b) Zeroth-order mode antenna constituted by left-handed transmission line [123]. (c) Effective ENZ antenna operating at the transverse cutoff mode of the substrate-integrated waveguide [124]. (d) Doped effective ENZ antenna with length-invariant operating frequency [125]. (e) High-gain single-point-fed antenna based on transmission-line grid with Dirac-type dispersion [131]. (f) Directivity enhancement via the load of index-near-zero superstrate [134]. Figures adapted with permission from: (a), [122], IEEE; (b), [123], IEEE; (c), [124], IEEE; (d), [125], IEEE; (e), [131], IEEE; (f), [134], IEEE.

Besides the length-independent operating frequencies, as a benefit in antenna design, another key property of the ZOR within ENZ metamaterial is the uniform distribution of fields in terms of both magnitude and phase. As a result, apertures composed by ENZ metamaterials can reach the ideally high aperture efficiency, indicating a high gain radiation. Aiming at achieving a high gain, antennas with electrically large dimensions based on ENZ metamaterials are also investigated. The wire medium and the metal grid, with ultralow costs and easy fabrication process, are two of the most popular choices to realize an electrically large ENZ medium while a line source or monopole antenna is most often used for feeding. A line source with infinite length within the wire medium has been proposed in [127], while a monopole embedded in a metal grid is used in [128], both of which illustrate that highly directive broadside radiation is achieved near the ENZ frequency. Following these theoretical and experimental results, a high-gain antenna based on a wire medium is proposed in [129]. In this antenna, four directive beams radiated by the ENZ medium in the azimuthal plane are all totally reflected to the broadside direction by a 45° interface between dielectric and air, thus generating a high-gain broadside sum-beam using a single feed. By coating the 45° interface with copper, a difference-beam is also available. With the progress of PCB technology, printed periodical structures are more and more widely used in ENZ antennas. One of the most typical designs is the Dirac leaky-wave antenna (LWA). For an ordinary LWA based on periodical structures, the beam scanning range is usually from backward to forward except for the broadside one due to the existence of open stopband, resulting from the coherent adding of the reflected waves by each unit. The ENZ metamaterial or metasurface is competent for solving this problem because of the Dirac-typed dispersion, in which the wave number is zero at the ENZ frequency. Using the ENZ metamaterial, a Dirac LWA is proposed in [130] for realizing a continuous beam steering from the backward to the forward direction. Fig. 4.1(e) shows a 2D Dirac LWA based on a transmission line grid to achieve a pencil beam with a gain of 18.8 dBi reported in [131]. Besides high-directivity applications, the uniformly distributed fields inside the whole ENZ material also indicate that the feeding point of the antenna can be arbitrarily chosen, enabling antennas that combine the impinging power from multiple input ports [132]. Another power combining method is through free space by enclosing several antennas with a MNZ shell [12]. This power combination is possible for realizing higher radiation power exceeding the limits of RF power amplifiers or front ends. On the topic of ENZ metamaterial antennas, the ENZ metamaterial not only is used as main resonators and radiators, but also can be utilized as loaded directors or reflectors for antennas. [133] demonstrates that an antenna horizontally placed on the interface of two different media tends to radiate more power into the medium with a lower wave impedance so that ENZ and MNZ materials are available as reflectors and directors, respectively. Fig. 4.1(f) depicts a MNZ superstrate loaded on a slot antenna for a 20-dB enhancement of the antenna's front-to-back ratio (FBR), which turns a bidirectional slot antenna into a directional one [134]. For higher bands like millimeter wave frequencies, this method also assists the gain enhancement of a circular-polarized antenna in [135] where a gain of 10 dBic is achieved. In [136], ENZ materials like ZnO are used as a reflective substrate for nanoantennas operating in near-infrared bands. As this discussion has made clear, the ENZ metamaterials provide new methods in antenna design, presenting unusual properties that can't be achieved by conventional antenna principles.

Generally, the ENZ metamaterial can contribute to a more uniform field distribution over the radiation apertures of the antennas and allow the operating frequency of the antenna to be insensitive to the variation of its size. On the other hand, the use of ENZ metamaterials could affect the bandwidth of the antenna, which usually leads to a narrower bandwidth of the antenna. The underlying physics is twofold. First, the ENZ metamaterials usually exhibit strong dispersion. Second, once the ENZ mode with uniform field distribution is excited over a large-size aperture of the antenna, electromagnetic power stored in and near the antenna would increase and thus result in a rise of the quality factor, which is defined as the stored electromagnetic power over the dissipation per circle.

4.2 ENZ Antennas Based on Photonic Doping

Similar to most of the ZOR antennas presented in the literature, the antenna proposed in [124] radiates a monopolar pattern because it generates two out-of-phase magnetic currents on two sides of the SIW. To achieve a broadside radiation, which is more widely used in communication systems, two in-phase magnetic currents are generated at the opening end of the SIW, inspired by the concept of photonic doping in [125], as exhibited in Fig. 4.2(a). The proposed antenna is composed based on a cutoff SIW, in which a ceramic block operating at its TM^z_{110} mode is inserted as a dopant to achieve a 180° phase shift. Under the excitation of a coaxial probe, a spatial static resonance is generated within the SIW as exhibited in Fig. 4.2(b). The magnitude of the electric field remains constant within the SIW other than the region occupied by the dopant. Unlike a normal patch antenna, which has a length-determined resonant frequency, the proposed antenna has an operating frequency insensible to the length of the SIW resulting from this uniform field distribution, as demonstrated in Fig. 4.2(c). The S-parameters shown in Fig. 4.2(c) indicate that the antenna maintains a good impedance matching at 5.5 GHz when its length varies from 0.22 λ_0 to 1.02 λ_0.

In contrast, unlike the length-insensible frequency, the antenna's radiation pattern is controlled by its length. To be specific, the underlying radiating mechanism of the antenna is illustrated as a two-element magnetic current

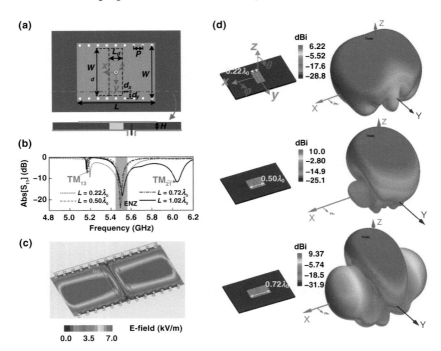

Figure 4.2 A doped ENZ antenna [125]. (a) The antenna's configuration from the top view and the side view. (b) The S-parameters in variation relative to the antenna's length L. (c) The magnitude distribution of the electric field within the doped equivalent ENZ medium. (d) Radiation patterns of the antenna with different L. (a)–(d) are adapted with permission from [125], IEEE.

array located at a distance of the length of the antenna L, similar to an ordinary patch antenna. Consequently, the antenna's radiation pattern and the broadside gain are both functions of L. Fig. 4.2(d) depicts the radiation patterns when L equals 0.22 λ_0, 0.5 λ_0, and 0.72 λ_0, respectively. The beamwidth decreases as L grows, and a sidelobe emerges when L exceeds 0.50 λ_0. A high directivity near 10.0 dBi is attained by optimizing L to 0.53 λ_0. It is necessary to mention that a normal patch antenna cannot realize such a high value because its length is supposed to be about half wavelength rather than an optimal one in terms of achieving the highest directivity.

A prototype of the antenna has been fabricated using PCB technique. The SIW is composed of a F4BMX substrate with a dielectric constant of 2.2 and a loss tangent of 0.002. Inside the SIW a rectangular hole is etched where a zirconia ceramic block with a dielectric constant of 34.0 and a loss tangent of 0.001 is tightly inserted. Finally, two copper sheets are soldered on both the top and bottom sides. For comparison and verification, three antenna prototypes with different lengths are assembled [120]. In these cases, L equals 0.22 λ_0, 0.5 λ_0, and 0.72 λ_0, respectively. The measured results show that all three prototypes have reflection coefficients lower than −10 dB at 5.5 GHz, and the measured radiation patterns match the simulations well.

4.3 ENZ Metamaterial Lenses and Absorbers

In this section, we focus on the transmissions and scatterings of slabs composed of ENZ metamaterial on propagating waves in free space and demonstrate the fascinating applications of achieving wave front manipulations using ENZ metamaterials, such as free-space beamforming as lenses and perfect absorption as absorbers, leading to enhanced performance for microwave applications. Three classes of examples are discussed and reviewed here: (1) wave front manipulation; (2) beam steering and imaging; (3) perfect absorbers with thin thickness.

To start with, the general scattering properties of a rectangular slab with zero refractive index studied by [16] are introduced. It illustrates by both analytical and numerical methods that whatever the incident wave front is, the transmitted wave front is always in-phase shaped by the ENZ slab. This phenomenon is generalized to an arbitrary-shaped ENZ slab in [132], demonstrating that the wave front is tailored to be conformal with the ENZ slab, which is utilized to isolate two regions of space and to tailor the phase pattern in one region, independent of incident phase distribution from the other region. Another issue raised by [137] is the discussion on transmission coefficients, in which it is shown that the reflections of oblique incidence are high for ENZ materials with only permittivity near zero as the phase matching condition. To enhance transmission, [138] and [139] both introduce losses and anisotropic permittivity to the ENZ material. Another efficient method is to design a matched ENZ metamaterial (i.e., the EMNZ material), with which a conceptual design of shaping the antenna's radiation pattern is presented in [140]. The tailoring of the radiation pattern using ENZ metamaterial is applied in various designs to obtain a high gain by generating a planar wave front. As shown in Fig. 4.3(a), a plano-concave EMNZ lens is realized by a bulk metal with densely etched air holes [141]. At 57.2 GHz, it transforms a spherical wave into a directive radiation. Compared with the proposed lens, directly covering the antenna's aperture with ENZ superstrate is a more space-saving approach. In [142], three layers of ENZ metamaterials are loaded above a patch antenna to realize an aperture efficiency as high as 80%. A similar idea is also workable for the gain enhancement of horn antennas. Usually, high-gain horn antennas demand large radiating apertures. In [143], by loading an ENZ lens on the aperture, 50% reduction of the horn's length is realized with no deterioration of the gain.

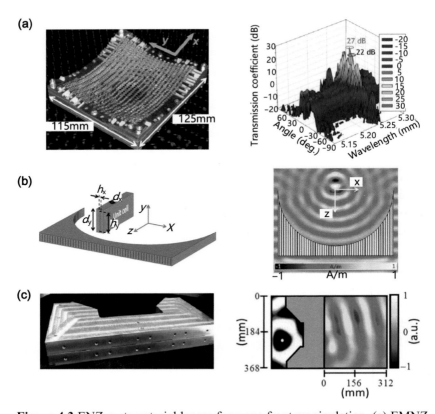

Figure 4.3 ENZ-metamaterial lenses for wave front manipulation. (a) EMNZ metamaterial lens for directivity enhancement [141]. Left: structure of the superlens constituted by a holey metallic metamaterial. Right: diagram of the angle- and wavelength-dependent realized gain of the superlens. (b) Waveguide-based ENZ lens to realize the Fourier transformation [144]. Left: ENZ metamaterial superlens made by waveguide array operating at the cutoff frequency. Right: transformation of a cylindrical wave front to a planar wave front via an ENZ metamaterial superlens. (c) ENZ metamaterial channel to generate arbitrary wave patterns [146]. Left: ENZ metamaterial channel with uniform slot array etched on the front face. Right: wave front transformation via the ENZ metamaterial channel as a lens. Figures adapted with permission from: (a), [141], APS; (b), [144], APS; (c), [146], IEEE.

As stated above, lenses composed of bulk ENZ metamaterials are capable of generating a high directive broadside radiation; however, beamforming seems to be a tough task since the uniform phase distribution is limited. To overcome this limitation for further powerful beam applications, an intuitive method is to cut the bulk ENZ lens into many channels along the propagating direction. A typical design is presented in Fig. 4.3(b), which cuts a plano-concave ENZ lens into numerous thin waveguides at the cutoff frequency [144]. The proposed lens is able to focus an obliquely impinging TM-polarized wave onto a single point, characterized by a spatial Fourier transformation from k-space to x-space on the focal plane. This idea is also used in [145] for 3D ENZ lens designs at 144 GHz with steered beam by moving the feeding horn. Furthermore, this method is also used to design ENZ lenses for TE-polarized incidence. As depicted in Fig. 4.3(c), an ENZ concave lens composed of four transversely placed cutoff waveguides can realize focusing of TE waves, according to [146], and reciprocally steering the radiated beams by moving the feeding monopole antenna in the focal plane [147]. Another way to realize focusing or beamforming is using anisotropic ENZ or MNZ metamaterials, which allows phase difference in one direction and demonstrates a uniformly distributed field in another one. Unlike the isotropic metamaterial, which shapes the wave front conformal with itself, [148] shows that the field in the anisotropic one has a wave front parallel to the axis along which the propagation constant is zero. Under an oblique incidence, the transverse phase distribution is maintained while the longitudinal phase difference is rectified, thus generating a steered beam with enhanced directivity. This property enables passive beam steering in millimeter waves by changing the loaded anisotropic ENZ superstrate over slot antennas [149]. Moreover, the hyperbolic metamaterial with zero permittivity along one axis supports the propagation of waves with high transverse wave number and assists subwavelength imaging [150]. In [151], an ENZ meta-lens based on waveguide structure was fabricated and validated in the terahertz frequency range.

ENZ slabs are also used to absorb incident waves with no reflections. It is analytically demonstrated in [152] that a grounded ENZ slab with an ultrathin profile obtains a perfect absorption for electromagnetic waves at a specific incident angle, enabling absorption based on metal and dielectric materials. Based on meander-line resonators, an ENZ absorber with a low profile of $\lambda_0/56$ is proposed with an absorptivity of 95% in [153]. Using periodical Swiss roll structures, another absorber is implemented with a height of $\lambda_0/80$ and a reflection coefficient lower than -10 dB for both polarizations [154]. The reflection has been lowered to -37 dB using CRLH as EMNZ metamaterial according to the most recent research [155]. All the absorbers mentioned in this section are composed of metals and dielectrics and use no lumped resistors, offering a low-cost solution for microwave absorber designs. Absorbers based on ENZ metamaterials also have applications in optical bands where ITO films are used for a zero permittivity at the plasmonic frequency. To obtain a broadband absorption, electric tuning is utilized as proposed in [156], in which an insulator layer is inserted between the ITO film and gold substrate. By adding a different voltage bias between the ITO and gold, the operating frequency is shifted in a wide range.

4.4 ENZ Metamaterial Transmission Lines and Circuit Devices

In this section, we investigate the applications of an ENZ channel as a key component in various kinds of microwave circuits. A thin channel filled with ENZ metamaterial or an arbitrarily shaped channel filled with EMNZ metamaterial exhibits a supercoupling effect under the excitation of a high-quality-factor ZOR mode, which inspires RF components including voltage-controlled couplers, RF filters, and high-sensitivity dielectric sensors. For circuit components, the cutoff waveguide is an important method for realizing ENZ metamaterials. Here, we also summarize three typical applications of microwave circuit components based on the ENZ metamaterials: (1) tunable couplers and high-sensitivity dielectric sensors; (2) multiband filters; (3) impedance-matching networks.

As the first example, the voltage-controlled coupler is usually realized by inserting electrically tuned capacitive loadings. In [157], a tunable channel is proposed with a voltage-controlled lumped inductor-capacitor (LC) resonator embedded in a thin ENZ waveguide. By changing the bias voltage on the varactor, the resonant frequency of the LC circuit is controlled, thus shifting the transmission band of the channel, obviously due to its high quality factor. Besides using lumped elements, a longitudinal slot is also utilized as a distributive capacitor, the influence of which is studied in [55] and [158]. Based on these results, a frequency-reconfigurable ENZ channel is proposed in [54] using two pin diodes as depicted in Fig. 4.4(a). A tunable range is realized from 6 GHz to 7 GHz, compensating the narrow band of traditional ENZ supercoupling. Moreover, mechanical tuning methods are also adopted with low cost and convenience for integration with multi-physics sensors. Based on the substrate-integrated impedance surface (SIIS) proposed in [159], periodic blind vias are used as a capacitive sheet, and a theoretical circuit model is derived to predict the performance of the SIIS, providing useful tuning ability for ENZ metamaterials. By implementing the SIIS structure of [159], the tunneling frequency of a thin ENZ channel is tuned in a wide range in [160] by mechanically controlling the insertion depth of the metal posts of SIIS. All the above designs offer components that can realize wideband ENZ supercoupling and new schematics for modulators. The fact that capacitive loading might have a huge influence on the tunneling frequency also provides the possibility of dielectric sensors with high precision and sensitivity. By placing a strip of the material under measurement, the frequency of the transmission peak is shifted corresponding to the permittivity of the perturbations, as proposed in [161]. The proposed permittivity sensor is convenient only for the solid dielectrics and has difficulty in measuring liquids. As shown in Fig. 4.4(b), a dielectric sensor for liquids is proposed in [162] by drilling a hole within the ENZ channel, which contains the sample to be tested. With enhanced electric

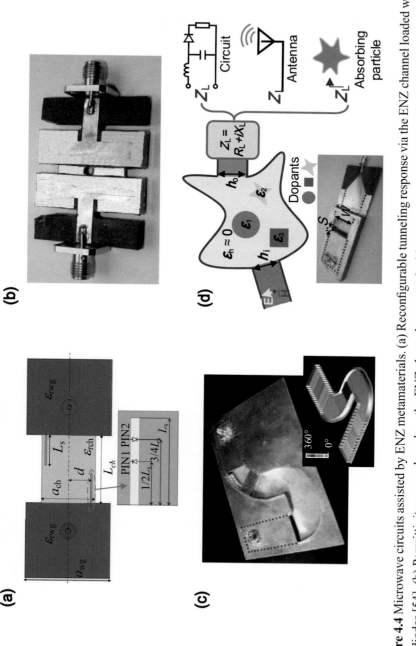

Figure 4.4 Microwave circuits assisted by ENZ metamaterials. (a) Reconfigurable tunneling response via the ENZ channel loaded with PIN diodes [54]. (b) Permittivity sensor based on the ENZ channel connected with two waveguides [162]. (c) Deformable waveguide transmission line made of EMNZ metamaterial [32]. (d) Conceptual illustration of general impedance circuit via a doped ENZ medium and the waveguide implementation of the one-stage general impedance-matching network [165]. Figures adapted with permission from: (a), [54], IEEE; (b), [162], IEEE; (c), [32], Creative Commons Attribution 4.0 International License; (d), [165], APS.

fields and higher quality factor, the ENZ dielectric sensor features a high sensitivity with respect to the permittivity of the sample.

As the second example, the ZOR mode in the ENZ channel is also adopted for cascaded cavity filter designs [53], [163], [164]. Chebyshev filters with two or three poles are proposed in [164] based on two or three cascaded ENZ channels weakly coupled with each other through a microstrip line. Besides the cascaded structure, both the ZOR and the Fabry–Perot (FP) modes can be coupled to generate a band-pass transmission. In [53] a second-order filter is designed using two thin channels, one of which operates at ZOR mode and the other at FP mode. To manipulate the resonant frequency of multiple modes for passband at the demanded frequency range, longitudinal slots are applied for ZOR mode, while inclined slots are etched on the channel under FP resonance. A dual-band filter is also proposed in [164] by connecting two resonant cavities through a thin channel that supports both ZOR and FP modes at different frequencies. The cavity's TM_{101} mode and the channel's ZOR mode combine to create a lower-frequency passband around 10.9 GHz, while the TM_{102} mode and the FP mode contribute to the higher one at 13.9 GHz.

Furthermore, a theoretical model of bulk ENZ metamaterial inserted in the waveguide transmission line is proposed in [165] for a general impedance-matching network, which is the third example in this section. An equivalent circuit model of a bounded ENZ metamaterial with arbitrary shape is built as a lumped serial inductor, which is proportional to the permeability of the enclosed ENZ medium. The general structure is inspired by the photonic doping method. As a result, a bulk ENZ metamaterial is modelled as a part of the circuit, even though huge geometrical discontinuities may exist in the ENZ channel. Based on this idea, as shown in Fig. 4.4(c), a deformable coupling channel using substrate-integrated, doped ENZ material is proposed in [32] to connect waveguides with different shapes with almost no reflections and phase shifts. This concept, referred to as "electric fiber," is useful and the material can be bent into any shape for on-chip interconnections. What's more, a general impedance-matching network is proposed using the ENZ metamaterial as a lumped inductor, of which the inductance is tuned by photonic doping for various loads including antennas, absorbers, and circuit components [165]. As can be seen from Fig. 4.4(d), the doped ENZ metamaterial transmission line component can eliminate the reflection between a thick waveguide and a thin one with obvious discontinuity. The transverse cavity operating at its cutoff frequency acts as an equivalent ENZ medium here, and the circuit model exactly describes its behavior, exhibiting a new path to design of flexible microwave devices and components with the geometry-irrelevant property of ENZ metamaterials.

4.5 ENZ Metamaterial Power Divider

The application of ENZ metamaterials is also extended to the design of multi-port components in microwave circuits. The theoretical analysis of multi-port ENZ hubs was presented in the pioneering works [166], [167]. As a concrete example, we subsequently demonstrate an ENZ-metamaterial-inspired multi-port power divider [168], which provides a uniform output phase distribution independent of the shape of the hub. The concept plot of a multi-port ENZ network comprising a photonic dopant is illustrated in Fig. 4.5(a). The ENZ metamaterial hub is branched to N waveguides, which are terminated by ports labeled as $P_1, P_2, \ldots, P_N$, respectively. Widths of the waveguides are denoted as $l_1, l_2, \ldots, l_N$, and wave impedances of materials filling the waveguides are $\eta_1, \eta_2, \ldots, \eta_N$. By imposing Faraday's law on the boundary of the ENZ metamaterial and invoking basic theories of microwave networks, the reflection coefficient at the port P_m is derived as

$$\mathbf{S}_{m,m} = \frac{j\omega\mu_0\mu_{\text{eff}}A + \sum_{p\neq m}\eta_p l_p - \eta_m l_m}{j\omega\mu_0\mu_{\text{eff}}A + \sum_{p\neq m}\eta_p l_p + \eta_m l_m}, \tag{4.1}$$

while the transmission coefficient from P_m to another port P_n $(n \neq m)$ can also be obtained:

$$\mathbf{S}_{n,m} = \frac{-2\sqrt{\eta_m\eta_n l_m l_n}}{j\omega\mu_0\mu_{eff}A + \sum_{p=1}^{N}\eta_p l_p}, n \neq m. \tag{4.2}$$

The scattering matrix of this N-port network of ENZ metamaterial can be organized in a compact form:

$$\mathbf{S}_{N\times N} = \mathbf{I}_{N\times N} - \frac{2}{j\omega\mu_0\mu_{\text{eff}}A + \sum_{p-1}^{N}\eta_p l_p}\xi_{N\times N}, \tag{4.3}$$

where $\mathbf{I}$ is an identity matrix, and ξ is a symmetric matrix with elements $\xi_{n,m} = (\eta_m\eta_n l_m l_n)^{1/2}$. Consider the EMNZ state (i.e., $\mu_{\text{eff}} \approx 0$). The ENZ metamaterial has a matched intrinsic impedance, and the zero-reflection condition for the input port ($\mathbf{S}_{1,1} = 0$) can be directly obtained from Eq. (4.1):

$$\sum_{p=2}^{N}\eta_p l_p = \eta_1 l_1. \tag{4.4}$$

Specifically, for the equal-split power divider, the output waveguides should have the same widths, and the scattering matrix takes the form

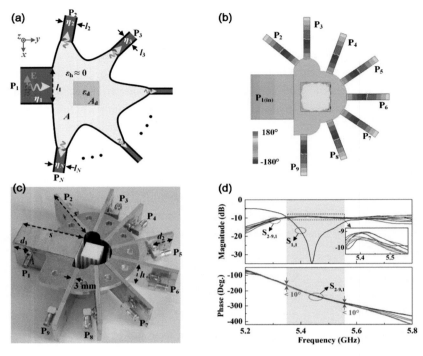

Figure 4.5 Multi-port power divider based on the ENZ metamaterial [168]. (a) Conceptual plot of 2D multi-port ENZ network. (b) Simulated magnetic field phase distribution of the proposed eight-way, equal-split metamaterial power divider. (c) Fabricated prototype of the proposed eight-way metamaterial power divider. (d) Measured transmission amplitudes and phases of the proposed power divider. (a)–(d) are adapted with permission from [168], IEEE.

$$\mathbf{S}_{N\times N} = \begin{bmatrix} 0 & \frac{-1}{\sqrt{N-1}} & \frac{-1}{\sqrt{N-1}} & \cdots & \frac{-1}{\sqrt{N-1}} \\ \frac{-1}{\sqrt{N-1}} & \frac{N-2}{N-1} & \frac{-1}{N-1} & \cdots & \frac{-1}{N-1} \\ \frac{-1}{\sqrt{N-1}} & \frac{-1}{N-1} & \frac{N-2}{N-1} & \ddots & \vdots \\ \vdots & \vdots & \ddots & \ddots & \frac{-1}{N-1} \\ \frac{-1}{\sqrt{N-1}} & \frac{-1}{N-1} & \cdots & \frac{-1}{N-1} & \frac{N-2}{N-1} \end{bmatrix}, \quad (4.5)$$

where the first element $\mathbf{S}_{1,1} = 0$ corresponds to the matched impedance seen at the input port P_1, and other elements in the first row or column of the scattering matrix represent the transmission coefficients. Following this guideline, an eight-way, equal-split power divider was designed [168], where a waveguide operating near the cutoff frequency of the TE_{10} mode is exploited to emulate the ENZ host. The simulated magnetic phase distribution is shown in Fig. 4.5(b), which clearly indicates a homogenous field configuration over the ENZ host and in-phase excitation of the output waveguides. Owing to the spatially static wave dynamics in the ENZ metamaterial, highly balanced allocation of power can be achieved via the hub with an arbitrary cross-sectional shape, which in turn substantially relaxes the requirement for strict geometrical symmetry of power dividers.

The fabricated proof-of-concept prototype of the eight-way, equal-split power divider is shown in Fig. 4.5(c). The hub cavity is manufactured via the metal machining technology, while the output waveguides are processed by the printed-circuit-board technology. The photonic dopant is made of Al_2O_3 ceramic, with a relative permittivity of 9.9. The measured amplitude and phase responses of the scattering coefficients are presented in Fig. 4.5(d). From 5.35 GHz to 5.56 GHz for the reflection coefficient $S_{1,1}$ smaller than −10 dB, the transmission amplitudes for eight output branches are uniformly around −9.6 dB with a maximum imbalance of 0.7 dB, while the maximum phase imbalance is only about 10°. Furthermore, by simply modifying the width of the output waveguides, the proposed scheme based on ENZ metamaterials can readily be extended to the design of power dividers with custom-designed division ratios.

5 Applications of ENZ Metamaterials in Optics and Quantum Physics

Compared with the components composed by ENZ metamaterials in microwave frequencies, the natural plasmonic materials with ENZ performances are critical in optical bands. For optical frequencies, the lack of high-conductivity metal and challenges in precise fabrication of complicated structures introduce considerable difficulties into the designing of optical devices. Therefore, ENZ metamaterials play an important role in optic applications, such as light trapping and confinement, nonlinear and nonreciprocal effect enhancement, and lumped optical circuits, also known as metatronics. ENZ metamaterials are also suggested as a promising platform to investigate and control rich quantum physics effects, ranging from the manipulation of the Purcell effect to boosting qubit entanglement. The essence of ENZ metamaterials applied to enhance classical and quantum optical effects may lie in the following aspects. First, a spatially uniform electromagnetic field enhancement can be achieved over the ENZ structure, which in turn boosts the local density of light as well as the light–matter interaction. Second, the homogenous phase configuration over the ENZ region strongly enhances the spatial coherence, which benefits the observation of collective emission effect and quantum entanglement effects. Additionally, the spatially static wave dynamics allows the non-radiation photonic state, which can be a powerful tool to modulate the radiation of optical sources.

5.1 Light Trapping

As the first example of optical application of ENZ metamaterials, we review trapping a photon for an infinitely long lifetime using a resonant cavity with an extremely large quality factor. For microwave bands, an air-filled resonant cavity fully enclosed by metal has a very high quality factor because metals (e.g., silver, copper, and aluminum) exhibit an ultrahigh conductivity close to that of PEC. However, the optical responses of metals suffer from high loss, while trapping photons in open structures might also induce the energy emission. Therefore, to realize a long lifetime resonance with a low decay rate is a tough task for optical band applications.

With the help of ENZ metamaterials, an optical resonance with infinite quality factor is possible, as stated in [169]. It is demonstrated that the core-shell structure supports a non-radiated mode where the electric and magnetic fields distribute only within the nanoparticle and are confined in the ENZ host. This unusual phenomenon is explained in Fig. 5.1(a) based on the mode analysis. Under each TM mode resonance of the nanoparticle, the magnetic fields on the interface between the particle and the ENZ host are forced to be zero. Furthermore, the homogeneity of magnetic fields within the ENZ medium ensures that the magnetic field at each position in the ENZ host is zero, which also applies to the boundary of the ENZ host. According to Huygens' principle, the radiated field of this structure is equivalent to a conformal PEC solid, the surface of which distributes currents characterized by $J_s = 2n \times H$, where n is the unit normal vector and H is the magnetic field. It shows that the radiation field of the proposed structure equals that of a PEC solid with no currents on the surface, which obviously equals zero. Nontrivially, as discussed in [43], the existence of such a non-radiating mode is irrelevant to either the size or the shape of the ENZ body surrounding the particle. Looking from the dielectric particle region into the ENZ region, the interface with the ENZ medium effectively acts as a perfect magnetic conductor boundary for the particle. Unlike ordinary resonant cavities whose eigenfrequencies are influenced by the geometry, this resonance presents a fixed resonant frequency under deformations [43] due to the utilization of the ENZ material. The excitation of this resonance is also discussed in [170], which demonstrates that this non-radiative mode is excited by a dipole placed at the center of the air bubble inside an ENZ medium. It also states that when

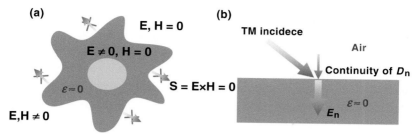

Figure 5.1 Demonstration of the light trapping effect and boundary effect of ENZ metamaterials. (a) Schematic demonstration of the mechanism underlying the 3D bound photonic state [43]. (b) Illustration of the enhancement of the normal component of the electric field in the ENZ region near the interface with air.

the dipole is placed off-center, a radiative mode would also be excited with the field distributions along the air–ENZ boundary. If the excited mode is not fulfilling the air–ENZ boundary, the energy cannot transmit to the ENZ host, leading to the non-radiative mode and noninteraction between two air bubbles. This resonance with no energy leakage is alternatively explained in [171] as a bound state in the radiation continuum, which is often found in the field of quantum physics, exhibiting connections in this interdisciplinary area. An optical meta-atom is constructed in [172] on this resonance with the functions of absorbing, trapping, storing, and emitting photons. The dielectric loss of the ENZ shell is compensated using gain medium, and the absorption and emission are enabled by the nonlinearity of the core. Trapping and storing light in a long lifetime may inspire new frontiers in optics including biological sensing, boosting of nonlinear optics, and quantum computing.

5.2 Boosting Optical Response of Matters

Nonlinear electronic devices are easily implemented in microwave bands using semiconductor elements, including diodes and transistors. However, for optical applications, to realize a nonlinear device for practical use is difficult because optical nonlinearity is naturally a weak effect in most materials so that high intensity is demanded to achieve an obvious nonlinear effect. The ENZ materials offer new chances to boost the optical nonlinearity of materials because of not only an enhanced electric field intensity but also new physical principles. It is also interesting that the electric field on the boundary of the ENZ material has a very large normal component and a zero tangential one simultaneously. As illustrated in Fig. 5.1(b), the TM wave obliquely impinges on the air–ENZ interface, which would lead to a significantly enhanced normal electric field in the ENZ medium due to the continuity of the normal electric displacement. In the normal medium, the linear permittivity is a finite number, while the nonlinear coefficient is very weak. As a result, the main part of the electric displacement is due to the linear one. In contrast, since the permittivity is zero, the major part of the electric displacement is contributed by the nonlinear term in the ENZ medium [173]. An analytical investigation of this effect is launched in [174], where a novel concept of nonlinear guided wave also is proposed, supporting the transverse power flow reversing effect, which means the power flow can have different signs along the transverse direction. By both boosting the nonlinearity and eliminating phase difference, the ENZ material is used for the generation of a bidirectional four-wave mixing laser, as depicted in Fig. 5.2 [22]. Since all the photons in the ENZ medium are all in-phase, the phase matching is not needed here, so both forward and backward lasers are generated. Bistability is another key nonlinear effect widely used in the optical area. A plasmonic silver slab loaded with periodic strips made of Kerr material proposed in [175] shows a bistable transmission, which enables tunable nonlinear devices. A significant application of the nonlinear effect is also used to generate higher-order harmonics of the input signal. The 3rd-, 5th-, 7th-, and 9th-order harmonics are all generated when an In-doped CdO (In:CdO) layer, performing at ultralow permittivity at 2.08 μm, is illuminated by a 2.08 μm laser [176]. A silicon-compatible design uses ITO as the ENZ material, as proposed in [177], where a 3rd-order harmonic is generated with an intensity proportional to the pump. Besides the odd-order ones, a 2nd-order harmonic is also generated in [178] using anisotropic ENZ metamaterials composed of one-dimensional layered Si and Dy:CdO materials. In addition, the enhanced nonlinearity also provides possibilities for optically tuned devices. The intensity of the pump laser determines a single-layered ITO's nonlinear refractive index, which also varies with incident angles [179]. Based on this tunable effect, an ultrafast all-optical switch is proposed in [180] with a responding time shorter than 1 ps under a low-power pump less than 4 mJ/cm^2. Aluminum-doped zinc oxide (AZO) is used here with a plasmon wavelength of 1.3 μm. Similar designs are presented in [181] with the use of ITO, and by using the ultrafast optical switch, a continuous wave is transferred into a femtosecond pulse train. Ultrafast all-optical switches are urgently needed in integrated photonic circuits [182].

Other optical effects, including nonlocal and nonreciprocal effects, are also enhanced using ENZ materials. The nonlocal dispersion of media yields a constitutive parameter depending on the wavevector of the electromagnetic wave, in other words, a spatial dispersion. Conventionally speaking, the spatial dispersion is a very weak effect in common media. However, this

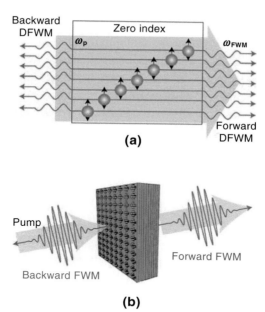

Figure 5.2 Enhanced optical nonlinearity via ENZ metamaterials.
Conceptual illustration (a) and metamaterial implementation (b) of the
zero-index-enabled phase matching for efficient four-wave mixing with equal
nonlinear generation in both forward and backward directions
[22]. This figure is adapted with permission from [22], AAAS.

effect is strongly boosted in artificial ENZ metamaterials. As a specific
example, a wire medium composed of a lattice of parallel metal wires, as
shown in Fig. 5.3(a), is used to emulate the ENZ medium [165]. According to
the analytical derivations in [183], this medium exhibits a unit permittivity
along the x and y directions, while along the z direction, a Drude-like
dispersive permittivity is obtained, which is related to the z component of
the wave vector. As a result, this artificial medium provides a strong spatial
dispersion together with nonlocal effects. The nonreciprocal effects of magneto-optic materials are also enhanced by ENZ metamaterials.
A nonreciprocal medium is characterized by an asymmetric constitutive
parameter tensor. In an ENZ medium, the diagonal elements approach zero
so that this asymmetry is enhanced. In [24], two different ways are proposed
to design such a nonreciprocal ENZ medium and an optical isolator, as
depicted in Fig. 5.3(b). The first method is to combine a magneto-optic
dielectric slab with positive permittivity and a metal slab with a negative
one, while the other one is filling a cutoff waveguide under TE_{10} mode with

the magneto-optic material. Both methods contribute to an artificial nonreciprocal ENZ medium providing the permittivity of an antisymmetric tensor with elements on the diagonal near zero. This structure is used for isolation of right-handed circular polarized (RCP) light, with the left-handed circular polarized (LCP) light as evanescent waves.

5.3 Metatronics and Other Optical Devices

As the last example of optical applications using ENZ metamaterials, the newly proposed technology of "metatronics" has brought the concept of lumped circuit from DC and RF regions to optical regions. The word "metatronics" consists of two parts: "meta" means metamaterials, and "tronics" means electronics and photonics. By combining these two terms together as "metatronics," we mean that by using the concept of metamaterials, we can merge the techniques of electronics and photonics, even though they are operating at different frequencies. Generally speaking, the concept of lumped circuits and elements, including resistors, inductors, and capacitors, applies only to the field of low-frequency electrodynamics up to RF bands with extremely small electric dimensions. The overall circuits are almost in-phase with "lumped" property. But optical devices usually have electrically large dimensions, such as mirrors, lens, and fibers, exhibiting a "distributed" property within the devices. Recently, based on the concept of metatronics, a series of research studies has implanted the classical concept of lumped circuits into optical bands by using metamaterials with zero or negative permittivity to realize an optical lumped circuit in nanoscale, also called optical metatronics [184].

As the general concept of metatronics, modeling the subwavelength plasmonic materials as lumped elements was first introduced in [33] in 2005. A schematic circuit model of such a nanoparticle with either positive or negative permittivity is shown in Fig. 5.4(a). With the illuminance of uniform planar waves, a nanoparticle with positive permittivity is modelled as a capacitor, while for that with a negative epsilon, the capacitor is replaced by an inductor. Therefore, the localized surface plasmon (LSP) can be modeled as LC resonance with plasmonic materials as inductors and the surrounding air as capacitors. What's more, it is the first time lumped inductors and capacitors with small electric dimensions (i.e., in deep subwavelength scale) and with an accurate theoretical model for nanocircuit design have been proposed. To avoid the coupling between two lumped elements in this paradigm, two ENZ slabs together with two epsilon-very-large (EVL) slabs compose a rectangular tank filled with plasmonic materials. The ENZ slabs are placed along the polarization of the incident wave, while the EVL slabs are perpendicular to it. Using this

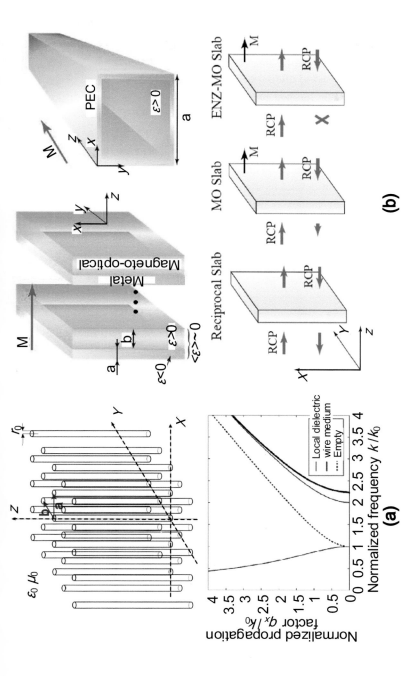

Figure 5.3 Boosting of nonlocal and nonreciprocal effects. (a) Boosting the nonlocal effect using a wire medium [183]. Top: The configuration of the wire medium. Bottom: The dispersion diagram. (b) Optical isolator based on boosted nonreciprocal effect by the ENZ medium [24]. Top: two different ways for realizing the ENZ magneto-optical medium. Bottom: The operating mechanism of the optical isolator. Figures adapted with permission from: (a), [183], APS; (b), [24], OSA.

method, the shielded nanoparticle performs like a lumped inductor or capacitor, depending only on the sign of its permittivity. Fig. 5.4(b) shows an example design of a metatronic circuit that operates like a lumped "DC" circuit under the incidence of TEM waves. In this paradigm of optical lumped circuits, the EVL medium has a function similar to that of conductors, and the ENZ medium functions as insulators to avoid mutual coupling between lumped elements.

After studying the basic lumped elements, here we focus on the shorting wires as the connection between lumped elements. For metatronic circuits, the conduit is needed to connect two elements without any magnitude or phase changes, even though it extends over a long distance or is bent into any shapes. As mentioned, the ENZ medium behaves as insulators. Therefore, the realization of such a nanowire in optical bands is based on the structure depicted in Fig. 5.4(c) where an air tunnel is drilled within a bulk ENZ medium, as discussed in detail in [35], [185]. Analogous to the conduction current $\mathbf{J_C} = \sigma\mathbf{E}$ in metallic wires for DC circuits, this air tunnel supports a constant electric displacement current $\mathbf{J_D} = \partial\mathbf{D}/\partial t$, thus it is called the D-dot wire. For metallic wires, the conductivity is quite large, and that of air is zero. Therefore, the

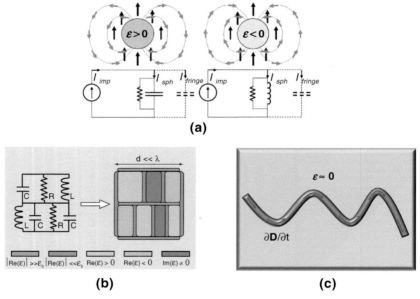

(a)

(b)   **(c)**

Figure 5.4 Optical metatronics. (a) Optical lump elements realized by nonplasmonic and plasmonic nanoparticles [33]. (b) Metatronic circuit tank consisting of nanoscale lumped elements and D-dot wire [34]. (c) D-dot wire embedded in the ENZ background as an optical conduit [186]. Figures are adapted with permission from: (a), [33], APS; (b), [34], AAAS.

conduction current is confined within the metallic wires without leakage to the air, even with sharp bending. But for displacement current, it is quite difficult to find special dielectric materials with large permittivity, and the permittivity of air is not zero. Therefore, the displacement current must leak to the air when the transmission dielectric is bent. For this reason, ENZ metamaterials are considered to take the place of air (i.e., we position the transmission dielectric inside the ENZ host to support the confinement of displacement current), operating as the D-dot wire. It is analytically demonstrated in [180] and experimentally verified in [186] at microwave frequencies that the phase shift of a D-dot wire is very small even if the wire is geometrically of several wavelengths. Based on this unique wire, a conceptual lumped-circuit board on which several nanoinductors and nanocapacitors are connected in a complex manner using the D-dot wire is proposed by [35]. As a summary of the concept of metatronics: in the new paradigm of optical "circuitry," the basic elements of inductors and capacitors are nanoparticles with negative and positive permittivity; the wires are air grooves or some other dielectric inside the ENZ host or background, supporting the displacement current as the information carrier with almost uniform phase within the circuits. The ENZ or ENG components are made of plasmonic materials, which may limit the optical metatronic circuit's design due to the intrinsic plasma frequency and large loss.

In addition to the implementation based on materials' properties, the metatronic circuit is also implemented by the metamaterials, which achieve zero or negative permittivities by the structural dispersion of a plate waveguide under its TE_{10} mode, as proposed in [37], offering a designable operating frequency by the height between two plates. According to the structural dispersion of the plate waveguide's TE mode, the stuffed dielectrics have an equivalent permittivity varying from negative to positive, and lumped inductors or capacitors are thus composed. Besides, the background dielectric acts as an exactly zero epsilon, so a D-dot wire is also implemented by a dielectric loop. This design is promising for metatronic circuits at any frequency that is independent of the material's intrinsic dispersions. Therefore, metatronic circuits are widely used to design optical devices and offer new design methods for nanophotonics. As a typical example, optical metatronic filters are available to be realized in subwavelength dimensions. By combining ENG and EPS nanoparticles along a transverse or longitudinal direction, a serial or parallel LC resonator is composed, respectively. This nanoresonator exhibits either a band-pass or a band-stop transmission for planar incident waves according to the circuit theory as designed in [187]. It demonstrates that such a circuit modeling method is very effective and offers new designing methodology for optical devices. It is noted that the effective waveguide inside the ENZ medium is infinite. Thus, an element with finite size can be treated as a lumped component without phase change. This is the reason that the optical metatronic circuit is embedded in the ENZ host to perform as a lumped circuit with uniform phase.

In the paradigm of metatronics, the ENZ medium operates as an insulator to confine the displacement current. In the work of [37], another paradigm of metatronics is also proposed by using the propagating electromagnetic wave as the information carrier instead of displacement current. For example, we can insert a metatronic LC element inside a waveguide at its propagation mode to control the power flow of the guiding wave to perform as a circuit. The element property is totally the same as the one with displacement current. Based on this, Butterworth filters of different orders are also available for optical bands using metatronics in [188] by determining the circuit topology and values of each element strictly following the designing guideline of a RF filter. And the concept of waveguide integrated circuits is also proposed in [189], with the Butterworth filter's design inside a propagation waveguide, exhibiting low insertion loss and cross talk, compared with the existing microstrip integrated circuits. What's more, a subwavelength-scaled ENZ block also can be modeled as lumped elements based on propagating mode. Therefore, another paradigm of metatronic circuits is proposed in [36], with different sets of lumped elements. As is illustrated in [36], there are four types of special zero-index materials that can be modeled as different lumped elements: (1) the ENZ medium with positive permeability operates as a series inductor, (2) the ENZ medium with negative permeability operates as a series capacitor, (3) the MNZ medium with positive permittivity operates as a shunt inductor, and (4) the MNZ medium with negative permittivity operates as a shunt capacitor. These results indicate that both series and shunt elements are available in optical form, based on the ENZ or MNZ metamaterials. In this way, we have further choices on the element selection inside.

Other optical devices based on the unusual index-near-zero effect were also proposed recently. A monochromatic multimode antenna operating at the infra-red region was designed based on the coupling with epsilon-near-zero Berreman mode [190]. An ultrafast, remotely triggered, all-optical switching was proposed and characterized based on the platform of ENZ nanocomposites [191], where the large third-order optical nonlinearity of ENZ multicomponent structure plays an important role in ensuring an ultralow threshold control intensity of light. Excitingly, [192] reported an ultracompact and unidirectional on-chip light source consisting of a paraboloid reflector etched in an aluminum-doped zinc oxide film, which exhibits ENZ response in the optical communication range. Another important class of optical applications results from the intriguing loss-induced effect in ENZ materials and metamaterials. In conventional materials, strong absorption usually requires that the material have either a high loss or a large thickness; however, the absorption by the zero-index medium absorbers exhibits an extraordinary behavior, that is, the required loss coefficient becomes negligibly small when the thickness of the zero-index medium absorbers tends to be zero [193]. Such a counterintuitive behavior can be explained via critical coupling to the wave propagation along the ENZ layer [193]. Then the unified theory framework for perfect absorption in ultra-thin films was established by Lai's group [194] under the condition of a constant electric or magnetic field across the films. Lai's group further extended the concept of coherent perfect absorption to enable a kind of geometry-irrelevant light absorber [111], [195], where the shape of the absorbers or the location of impurities would hardly affect the wave absorption. The shape- or position-irrelevant characteristics originate from the uniformly distributed fields over the zero-index region. The loss-induced effect in index-near-zero metamaterials was also exploited to bend the direction of electromagnetic power flow [196], for field enhancement, and for light collimation [197], [90].

5.4 Applications in Quantum Physics and Other Fields

Up to this point we have thoroughly discussed the applications of ENZ metamaterials in the classical optics regime, where the physical sizes of the systems involved are comparable with the wavelength. Driven by the rapid development of nanoscience and technology, a wide range of fields, including materials science and electronic engineering, are moving to smaller and smaller scales, such as that of a molecule or an atom. In this scenario, classical optics is insufficient to explain the emergent physics effects – we are entering the regime of quantum physics, where the states of microscopic particles as well as their interaction with waves have to be described in a quantized approach. This section presents the exciting applications of ENZ metamaterials in tailoring wave–matter interaction from the perspective of quantum physics.

In the seminal work by Alù et al. [198], the ENZ metamaterial, made of a hollow waveguide channel operating near the cutoff frequency, is demonstrated to significantly boost molecular fluorescence. The enhancement of spontaneous emission for a molecule or an atom coupled with external electromagnetic resonance was first studied by Purcell and was known as the Purcell effect. Traditional schemes proposed to boost spontaneous emission via the use of resonant cavities or plasmonic resonators were limited by sensitive dependence on the location of microscopic particles. The waveguide-based ENZ metamaterial, on the contrary, can provide a uniform electric field enhancement, which in turn allows huge emission enhancement of a molecule independent of its position along the waveguide channel. Later, Fleury et al. unveiled that ENZ metamaterials are also capable of boosting cooperative spontaneous emission [25]. The collective emission of N collaborating quantum emitters is known as the superradiance effect, where the radiation intensity is expected to be proportional to N^2 rather than N. The key to ENZ metamaterials' ability to realize superradiance enhancement lies in the extremely stretched wavelength, which increases the cooperation length of a resonant system and thus ensures that arbitrarily distributed quantum emitters operate coherently. The waveguide structure to emulate ENZ response is illustrated in Fig. 5.5(a), where a quantum emitter with a dipole moment d is placed in the ENZ channel. The calculated Purcell factor (i.e., the spontaneous emission rate γ over its value γ_0 in a vacuum) is presented in Fig. 5.5(b), which clearly validates a position-independent enhancement of spontaneous emission near 400 THz, the predesigned ENZ frequency of the waveguide. For the case of collective emission of multiple quantum emitters, the calculated maximum emission gain is shown in Fig. 5.5(c), where a strong enhancement of collective radiation is observed near the ENZ frequency.

Sokhoyan et al. [199] elucidated the enhanced Purcell factor in the ENZ waveguide from the perspective of increased local density of optical states, which is defined as the number of photonic modes per unit frequency and unit volume at the position under consideration. The ENZ waveguide with tailored local density of optical states can efficiently modify the spontaneous emission rate of an emitter. As another important concept in quantum electrodynamics, the Lamb shift reflects the extremely weak coupling of the bound electron to vacuum modes. Introducing inhomogeneous scatterers near the emitter is a feasible approach to modify the value of Lamb shift. Inspired by that, artificial structures such as photonic crystals and plasmonic structures have been exploited to engineer the scattering Green's function and thus to amplify the Lamb shift effect. The work [199] also theoretically demonstrated that, by placing the emitter in the center of a rectangular waveguide operating at the cutoff frequency, the value of the Lamb shift should be two orders of magnitude larger than that in free space.

To realize fine control of the quantum emission, more sophisticated ENZ meta-structures are proposed. In the work of Liberal et al. [170], an open ENZ cavity containing a vacuum bubble was designed, and by manipulating the displacement of a quantum emitter from the center of the bubble, one can conveniently switch the non-radiating and radiating modes of the emitter. The physical mechanism underlying such an unusual effect is briefly explained here. The non-radiating mode of a quantum emitter is actually related to the

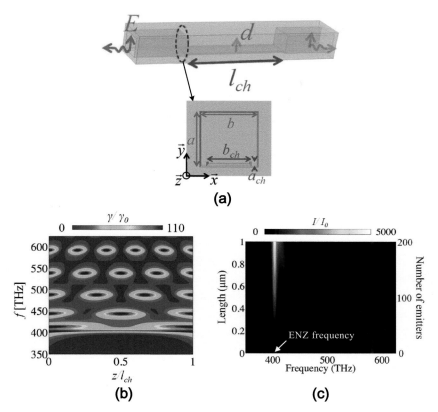

Figure 5.5 Quantum emission assisted by ENZ metamaterial. (a) Configuration of the narrow plasmonic channel connecting two rectangular waveguides. (b) Spontaneous emission rate versus frequency and position along the channel. (c) Maximum emission gain compared to the same number of molecules in free space. Figures (a), (b), and (c) are adapted with permission from [25], APS.

excitation of the zero-magnetic-field mode in the ENZ body, and the electric field distribution is spatially static. Such a field configuration yields a vanishing Poynting vector outside the ENZ cavity and thus inhibits the radiation of the quantum emitter in the bubble. Changing the position of the emitter can break the zero-magnetic-field mode and lead to a nonzero radiation to the environment.

The works we have discussed in this section have elucidated how to increase the coupling of a quantum emitter to the electromagnetic modes in a vacuum. On the other hand, the ENZ metamaterial can also be exploited to suppress the vacuum modes, alternatively, the vacuum fluctuation. As discussed in [200], the density of states $\rho(\omega)$ in a space filled with a homogenous medium of refractive index n is $n^3(\omega^2/\pi^2 c^3)$, which would completely vanish as n approaches zero. Based on this principle, an EMNZ shell is designed in [200], and a dipole emitter is located inside the index-near-zero shell. Compared with a dielectric shell, an EMNZ shell efficiently shields the inner core from the vacuum electric field fluctuation. Other interesting applications of ENZ metamaterials in quantum optics include long-distance quantum entanglement assisted by ENZ films [201], ENZ-metamaterial-based grid structures serving as on-chip quantum networks [202], and so on. In a nutshell, ENZ metamaterials are providing new horizons for research in the quantum physics regime and have demonstrated huge potential in quantum effects enhancement and manipulation, quantum information processing, and other applications.

Besides having special functions in classical and quantum optics, ENZ meta-materials could impact a range of multiphysics applications. The work by Rodríguez-Fortuño et al. [203] presented an intriguing mechanical effect in which the ENZ metamaterial can exert a repulsive force on an electric dipole placed near its surface, which can be a high-frequency analogy to the Meissner effect of a semiconductor with perfect diamagnetism. Rigorous theoretical and numerical studies were performed based on the Maxwell stress tensor, which shed light on the fact that the radiation pressure acting on the dipole source would repel it away from the ENZ body. Such a repulsion effect is also shown to be insensitive to the loss and allowed to be achieved at an arbitrary frequency, thus providing new opportunities for creating micromechanical devices. ENZ metamaterials were also harnessed to engineer the thermal radiation [204]. The spatially static character of the fields in the ENZ body leads to a strong spatial coherence, which is beneficial for directive thermal emission. Such a field configuration is not tied to the geometry of ENZ metamaterials, and we are empowered to manipulate flexibly the pattern of the thermal emission. It is noted that there will be a compromise between the directivity of emission pattern and the amount of the thermal emission power. As discussed in [204], while the strength of the fluctuat-ing current source is proportional to the material loss, increase of the loss can be detrimental to the spatial coherence of the fields in ENZ media and thus deviate the emission directivity from the optimal value in the ENZ limit. The concept of near-zero index also inspires advances in thermal metamaterial cloaking [14], where an interesting correspondence between the zero index in Maxwell's equa-tions and the infinite thermal conductivity in Fourier's law is identified. Recently, an analogy between the index-near-zero media and the ideal fluids has been found [205], as the obstacles embedded in index-near-zero media would not incur any vortex or "turbulence" in Poynting vector distribution, which resembles the case of ideal fluids. Consequently, index-near-zero metamaterials will serve as new platforms for research in fluid mechanics.

Another important and promising research direction is the combination of ENZ/EMNZ media with other nanophotonics techniques. One important example is the combination of zero-index materials with metasurfaces to realize the famous cloaking phenomenon. For example, in [206], by combining the wave front tailoring functionality of transparent metasurfaces and the wave tunneling effect of zero-index materials, a unique type of hybrid invisibility cloak with transmission-type geometry was realized. A rhombic double-layer cloaking shell was constructed for experimental verification, which consisted of a nearly transparent metasurface and an EMNZ medium composed of dielectric photonic crystals with Dirac-cone dispersions. This is an important development because the cloaking effect was previously possible only through transformation optics or scattering cancellation at low frequencies, while the combination of EMNZ media and metasurfaces provides a fundamentally new route. Another important example of this direction is the combination of ENZ media with PT-symmetric metasurfaces, which has been discussed in [112], where the extraordinary physical property of impurity-immunity was realized when a slab of ENZ medium was sandwiched by a pair of parity-time (PT)-symmetric metasurfaces. The two complementary solutions of exceptional points of the system coalesce into one when the permittivity of the slab approaches zero. As a result, the doping effect of the impurities in ENZ media can be eliminated.

6 Summary and Outlook

6.1 Summary of This Element

The ENZ metamaterials have been experiencing rapid development in the past decades and have exerted far-reaching impacts on a wide range of fields in science and engineering. In this short Element, we have elucidated the concepts of the ENZ metamaterials along with the more generalized index-near-zero metamaterials by identifying their special coordinates in the ε-μ phase space. Starting from there, we systematically reviewed the unusual effects arising from the vanishing optical index, including wave dynamics with spatially static features, geometry-independent wave properties, and anomalous tunneling of power (i.e., the super-coupling effect). Then, various approaches to realize or emulate ENZ response were reviewed, ranging from the naturally occurring plasmonic materials to artificially structured materials. Therein, ENZ metamaterials with periodic and nonperiodic styles were demonstrated with abundant design examples. It should be remarked here that ENZ metamaterials empowered with the technique of

photonic doping have established a paradigm of spatial-order-irrelevant metamaterials, which never could have been expected in conventional single or double
negative metamaterials and thus will be of significance to the basic theoretical
framework of artificially structured media. As a central part of this Element, we
demonstrated the exciting applications of ENZ metamaterials in microwave
engineering, optics, quantum physics, and other multi-physics fields. ENZ metamaterials have been exploited to manipulate wave–matter interactions in both
microscopic and macroscopic scales, creating opportunities to boost weak optical
and quantum effects. Besides providing fertile ground for fundamental research
in physics and materials science, ENZ metamaterials are shaping a new platform
for wave engineering with high degrees of freedom, with practical applications in
antenna and circuit designs.

It is quite interesting to note that ENZ and NZI metamaterials, while occupy-
ing only a narrow region in ε-μ phase space, do open a broad horizon for
research in the regime of artificially structured media. The essence of ENZ
metamaterials lies in two aspects. First, the infinitely stretched wavelength
equates an electrically large space to a single "point", which ensures
a uniform field distribution and eliminates the dependence of ENZ devices on
their detail geometries. Second, as every point in the ENZ region is electromagnetically identical, tuning at any point can sufficiently impact the response of
the entire system, which gives access to the manipulation of electromagnetic
waves with an ultrahigh sensitivity. It is worth mentioning that the concepts,
analysis methods, and metamaterial design philosophies introduced in this
Element can be, in principle, applied to a wide range of wavelengths, and can
even inspire the study of index-near-zero metamaterials in other fields such as
acoustics and thermology.

6.2 Challenges and Outlook for Future Directions

At the end of this Element, we note the challenges encountered in the development of ENZ metamaterials and represent an outlook for the promising directions in this field. The metamaterial-based approach has been developed to
realize the ENZ response at microwave, terahertz, and optical regions.
However, a significant bottleneck for the implementation of ENZ metamaterials
or plasmonic media is the extremely narrow bandwidth. Consider an ENZ
metamaterial whose relative effective permittivity is described by a lossless
Drude model: $\varepsilon_r(\omega) = 1 - \omega_p^2/\omega^2$, the frequency range for $|\varepsilon_r| < 0.01$ is
approximately from $-0.5\%\omega_p$ to $0.5\%\omega_p$, that is, only 1% fractional bandwidth
for ENZ response. The ENZ phenomenon yielded by the strong resonance of
metamaterial elements can face even narrower operating bandwidth. In many

practical scenarios, such as communication and signal processing, bandwidth is a key factor to be considered, and excessively narrow bandwidth would severely limit the system performance. Hence, broadening the bandwidth of ENZ and NZI metamaterials can be of crucial significance for their real-world applications. Generally, there could be two approaches to attain the ENZ and NZI response with relatively larger bandwidth. The first approach is to reduce the quality factor of the resonance of the metamaterial elements. For example, in the context of photonic doping, the use of dielectric impurities with larger volume and lower permittivity can efficiently reduce the spectral quality factor, and thus lead to a flatter dispersion of the effective optical index near the EMNZ response. Following a similar line of reasoning, the scheme based on nonresonant dielectric constituent elements usually yields a larger bandwidth than those of metallic-resonator-based metamaterials. The second approach to increase bandwidth is to introduce multiple closely spaced zeros in the dispersive effective permittivity and/or permeability. The studies by Sun et al. [73], [207], [208] employed the Milton representation and the Bergman representation to design permittivity functions with multiple cross-zero points. This idea was practically implemented by a multilayer stacked structure [73], where each layer contains a dielectric host and differently sized metallic inclusions. It was shown that, by altering the volume fractions of metallic inclusions and the thickness of each layer, one can manipulate the zeros of the effective permittivity. Recalling Eq. (2.4) and imposing the ENZ condition ($\varepsilon_r \ll 1$, and $\mu_r = 1$), the expression of group velocity in the ENZ limit reduces to $v_g = 2c\sqrt{\varepsilon_r}/(\omega \cdot \partial\varepsilon_r/\partial\omega)$. As seen, a weaker dispersion of permittivity would lead to a larger group velocity, which, however, has an upper limit of c so as to obey the principle of relativity. In substance, ENZ media should always be dispersive, which in turn imposes a limit to their bandwidth. It should be mentioned that both natural and engineered permittivity functions should follow the Kramers–Kronig relationship [209], as required from the causality of a realizable linear system. Roughly speaking, it is impossible to realize an extreme wideband nondispersive ENZ response without the imaginary part of the permittivity function.

Another challenge that ENZ and NZI metamaterials face is the problem of losses, which is actually a common issue for plasmonic materials and metamaterials. Generally, depending on the underlying physical mechanisms, losses can be roughly categorized into two classes. The first class of loss is due to the imperfections of materials, such as ohmic dissipation and dielectric loss, which can arise from collision of free carriers or the interband transitions of electrons [62]. The second kind of loss is due to the leakage of waves from artificially engineered structures, which is manifested as unwanted electromagnetic coupling or radiation. In the context of ENZ metamaterials, the enhanced quality factor would exaggerate the detrimental effects of losses on the performance of the ENZ-metamaterial-based devices. To deal with the plasmonic losses, one can consider using low-loss plasmonic materials, such as the noble metals in the optical region and transparent conductive oxides in the infrared region. Additionally, some noble metals can also serve as good conductors with small ohmic loss in the radio frequency region. To combat the leakage loss, we have to expend more efforts on the metamaterial design in structure level. Among the approaches to realize ENZ response, the waveguide-based metamaterials, characterized by fully closed structures, have been evidenced to yield a much smaller leakage loss. Additionally, it was shown by Dong et al. [88], via suitably adjusting the thickness of the photonic crystal, that the upward and downward out-of-plane radiation can be destructively interfered, which leads to a bound photonic state in continuum. As an advance in the realization of low-loss plasmonic material, the work of Li et al. [71] unveiled the possibility of exploiting the structural dispersion of waveguide to reduce the losses in the ENZ and plasmonic polariton waves. The essential idea of this work is that, for a lossy plasmonic medium filled in a metallic waveguide, the zero for the real part of its effective permittivity under the TE_{10}-mode operation would move to a low-loss region. This work provides a promising route to reduce the loss of ENZ metamaterials from both the structural and material perspectives.

In addition to pursuing ENZ response with wideband and low-loss properties, there will be a growing tendency to design ENZ metamaterials allowing for on-chip integration. As required by the miniaturization and easy interconnection of circuit modules, recent decades have witnessed a strong trend for photonic and electronic circuitries to be integrated in a planar platform. To accommodate ENZ metamaterials on electronic and photonic chips, one should consider different fabrication techniques dependent on different frequency regions. For example, the techniques of printed circuit board and substrate integrated waveguide are suitable for metamaterial implementation in microwave- and millimeter-wave regions, while the bulk silicon micromachining technique and nanofabrication technique are respectively suitable for implementation in the terahertz and optical regions. Among different approaches to achieve ENZ response, the waveguide-based ENZ metamaterials, with low leakage loss and low mutual coupling characteristics, will have special advantages in maintaining high signal integrity and thus will allow a dense integration with planar circuits. Another exciting direction is the combination of ENZ metamaterials with emergent machine learning technology. The development of artificial intelligence has led to a revolution in information technologies, radically changing the methods of data mining, processing, and understanding. On the other hand, ENZ and NZI metamaterials have opened unprecedented degrees of freedom to manipulate the electromagnetic wave – the carrier of information – and hence ENZ and NZI metamaterials are promising to serve as a physical platform to implement the advanced algorithms used in artificial intelligence. Furthermore, the metamaterial-based information processing can even fuel the rise of photonic computers [210], which enable computation at the speed of light and significantly improved parallelism.

References

1. S. Zouhdi, A. Sihvola, and M. Arsalane, Advances in Electromagnetics of Complex Media and Metamaterials. Dordrecht: Springer Netherlands, 2002.
2. R. Marqués, F. Martín, and M. Sorolla, Metamaterials with Negative Parameters. Hoboken, NJ: John Wiley & Sons, Inc., 2007.
3. C. Caloz and T. Itoh, Electromagnetic Metamaterials: Transmission Line Theory and Microwave Applications. Hoboken, NJ: John Wiley & Sons, Inc., 2005.
4. W. Cai and V. Shalaev, Optical Metamaterials. New York: Springer, 2010.
5. T. J. Cui, D. R. Smith, and R. Liu, Metamaterials. Boston, MA: Springer, 2010.
6. N. Engheta and R. W. Ziolkowski, Metamaterials. Hoboken, NJ: John Wiley & Sons, 2006.
7. N. I. Zheludev and Y. S. Kivshar, From metamaterials to metadevices, Nat. Mater., 11, 11, 917–924 (2012).
8. A. M. Shaltout, N. Kinsey, J. Kim et al., Development of Optical Metasurfaces: Emerging Concepts and New Materials, Proc. IEEE, 104, 12, 2270–2287 (2016).
9. I. Liberal and N. Engheta, Near-zero refractive index photonics, Nat. Photonics, 11, 3, 149–158 (2017).
10. O. Reshef, I. De Leon, M. Z. Alam, and R. W. Boyd, Nonlinear optical effects in epsilon-near-zero media, Nat. Rev. Mater., 4, 8, 535–551 (2019).
11. X. Niu, X. Hu, S. Chu, and Q. Gong, Epsilon-near-zero photonics: a new platform for integrated devices, Adv. Opt. Mater., 6, 10, 1–36 (2018).
12. Q. Cheng, W. X. Jiang, and T. J. Cui, Spatial power combination for omnidirectional radiation via anisotropic metamaterials, Phys. Rev. Lett., 108, 21, 2–6 (2012).
13. F. Liu, X. Huang, and C. T. Chan, Dirac cones at $k \rightarrow = 0$ in acoustic crystals and zero refractive index acoustic materials, Appl. Phys. Lett., 100, 7 (2012).
14. Y. Li, K. Zhu, Y. Peng et al., Thermal meta-device in analogue of zero-index photonics, Nat. Mater., 18, 1, 48–54 (2019).
15. N. Kinsey, C. DeVault, A. Boltasseva, and V. M. Shalaev, Near-zero-index materials for photonics, Nat. Rev. Mater., 4, 12, 742–760 (2019).
16. R. W. Ziolkowski, Propagation in and scattering from a matched metamaterial having a zero index of refraction, Phys. Rev. E, 70, 4, 046608 (2004).

17. M. Silveirinha and N. Engheta, Tunneling of electromagnetic energy through subwavelength channels and bends using ε-near-zero materials, Phys. Rev. Lett., 97, 15, 157403 (2006).

18. B. Edwards, A. Alù, M. E. Young, M. Silveirinha, and N. Engheta, Experimental verification of epsilon-near-zero metamaterial coupling and energy squeezing using a microwave waveguide, Phys. Rev. Lett., 100, 3, 033903 (2008).

19. R. Liu, Q. Cheng. T. Hand et al., Experimental demonstration of electromagnetic tunneling through an epsilon-near-zero metamaterial at microwave frequencies, Phys. Rev. Lett., 100, 2, 023903 (2008).

20. S. Enoch, G. Tayeb, P. Sabouroux, N. Guérin, and P. Vincent, A Metamaterial for Directive Emission, Phys. Rev. Lett., 89, 21, 213902 (2002).

21. A. Ciattoni, C. Rizza, and E. Palange, Extreme nonlinear electrodynamics in metamaterials with very small linear dielectric permittivity, Phys. Rev. A, 81, 4, 043839 (2010).

22. H. Suchowski, K. O'Brien, Z. J. Wong et al., Phase mismatch-free nonlinear propagation in optical zero-index materials, Science, 342, 6163, 1223–1226 (2013).

23. R. J. Pollard, A. Murphy, W. R. Hendren et al., Optical nonlocalities and additional waves in epsilon-near-zero metamaterials, Phys. Rev. Lett., 102, 12, 127405 (2009).

24. A. R. Davoyan, A. M. Mahmoud, and N. Engheta, Optical isolation with epsilon-near-zero metamaterials, Opt. Express, 21, 3, 3279 (2013).

25. R. Fleury and A. Alù, Enhanced superradiance in epsilon-near-zero plasmonic channels, Phys. Rev. B, 87, 20, 201101 (2013).

26. W. Rotman, Plasma simulation by artificial dielectrics and parallel-plate media, IRE Trans. Antennas Propag., 10, 1, 82–95 (1962).

27. A. Sanada, M. Kimura, I. Awai, C. Caloz, and T. Itoh, A planar zeroth-order resonator antenna using a left-handed transmission line, Conf. Proceedings- Eur. Microw. Conf., 3, 1341–1344 (2004).

28. P. Moitra, Y. Yang, Z. Anderson et al., Realization of an all-dielectric zero-index optical metamaterial, Nat. Photonics, 7, 10, 791–795 (2013).

29. X. Huang, Y. Lai, Z. H. Hang, H. Zheng, and C. T. Chan, Dirac cones induced by accidental degeneracy in photonic crystals and zero-refractive-index materials, Nat. Mater., 10, 8, 582–586 (2011).

30. Y. Li, S. Kita, P. Muñoz et al., On-chip zero-index metamaterials, Nat. Photonics, 9, 11, 738–742 (2015).

31. I. Liberal, A. M. Mahmoud, Y. Li, B. Edwards, and N. Engheta, Photonic doping of epsilon-near-zero media, Science, 355, 6329, 1058–1062 (2017).

32. Z. Zhou, Y. Li, H. Li et al., Substrate-integrated photonic doping for near-zero-index devices, Nat. Commun., 10, 1, 4132 (2019).

33. N. Engheta, A. Salandrino, and A. Alù, Circuit elements at optical frequencies: nanoinductors, nanocapacitors, and nanoresistors, Phys. Rev. Lett., 95, 9, 095504 (2005).

34. N. Engheta, Circuits with light at nanoscales: optical nanocircuits inspired by metamaterials, Science, 317, 5845, 1698–1702 (2007).

35. A. Alù and N. Engheta, All optical metamaterial circuit board at the nanoscale, Phys. Rev. Lett., 103, 14, 143902 (2009).

36. F. Abbasi and N. Engheta, Roles of epsilon-near-zero (ENZ) and mu-near-zero (MNZ) materials in optical metatronic circuit networks, Opt. Express, 22, 21, 25109 (2014).

37. Y. Li, I. Liberal, C. Della Giovampaola, and N. Engheta, Waveguide metatronics: lumped circuitry based on structural dispersion, Sci. Adv., 2, 6, e1501790 (2016).

38. J. B. Khurgin and A. Boltasseva, Reflecting upon the losses in plasmonics and metamaterials, MRS Bull., 37, 8, 768–779 (2012).

39. J. B. Khurgin, How to deal with the loss in plasmonics and metamaterials, Nat. Nanotechnol., 10, 1, 2–6 (2015).

40. P. B. Johnson and R. W. Christy, Optical constants of the noble metals, Phys. Rev. B, 6, 12, 4370–4379 (1972).

41. M. Silveirinha and N. Engheta, Design of matched zero-index metamaterials using nonmagnetic inclusions in epsilon-near-zero media, Phys. Rev. B, 75, 7, 075119 (2007).

42. A. M. Mahmoud and N. Engheta, Wave–matter interactions in epsilon-and-mu-near-zero structures, Nat. Commun., 5, 1, 5638 (2014).

43. I. Liberal, A. M. Mahmoud, and N. Engheta, Geometry-invariant resonant cavities, Nat. Commun., 7, 1, 10989 (2016).

44. F. J. Garcia-Vidal, L. Martin-Moreno, T. W. Ebbesen, and L. Kuipers, Light passing through subwavelength apertures, Rev. Mod. Phys., 82, 1, 729–787 (2010).

45. G. Castaldi, I. Gallina, V. Galdi, A. Alù, and N. Engheta, Electromagnetic tunneling through a single-negative slab paired with a double-positive bilayer, Phys. Rev. B, 83, 8, 081105 (2011).

46. M. G. Silveirinha and N. Engheta, Theory of supercoupling, squeezing wave energy, and field confinement in narrow channels and tight bends using ε near-zero metamaterials, Phys. Rev. B, 76, 24, 245109 (2007).

47. A. Alù, M. G. Silveirinha, and N. Engheta, Transmission-line analysis of ε-Near-Zero-filled narrow channels, Phys. Rev. E, 78, 1, 016604 (2008).

48. J. S. Marcos, M. G. Silveirinha, and N. Engheta, μ -near-zero supercoupling, Phys. Rev. B, 91, 19, 195112 (2015).
49. H. Feng Ma, J. Hui Shi, Q. Cheng, and T. Jun Cui, Experimental verification of supercoupling and cloaking using mu-near-zero materials based on a waveguide, Appl. Phys. Lett., 103, 2, 021908 (2013).
50. A. Alù and N. Engheta, Light squeezing through arbitrarily shaped plasmonic channels and sharp bends, Phys. Rev. B, 78, 3, 035440 (2008).
51. B. Edwards, A. Alù, M. G. Silveirinha, and N. Engheta, Reflectionless sharp bends and corners in waveguides using epsilon-near-zero effects, J. Appl. Phys., 105, 4, 044905 (2009).
52. L. Liu, C. Hu, Z. Zhao, and X. Luo, Multi-passband tunneling effect in multilayered Epsilon-Near-Zero Metamaterials, Opt. Express, 17, 14, 12183 (2009).
53. N. Vojnovic, B. Jokanovic, M. Radovanovic, and F. Mesa, Tunable second-order bandpass filter based on dual ENZ waveguide, 2015 9th Int. Congr. Adv. Electromagn. Mater. Microwaves Opt. METAMATERIALS 2015, September, 316–318 (2015).
54. M. Mitrovic, B. Jokanovic, and N. Vojnovic, Wideband tuning of the tunneling frequency in a narrowed epsilon-near-zero channel, IEEE Antennas Wirel. Propag. Lett., 12, 631–634 (2013).
55. N. Vojnovic, B. Jokanovic, M. Radovanovic, F. Medina, and F. Mesa, Modeling of nonresonant longitudinal and inclined slots for resonance tuning in ENZ waveguide structures, IEEE Trans. Antennas Propag., 63, 11, 5107–5113 (2015).
56. Y. Liu, F. Sun, Y. Yang et al., Broadband electromagnetic wave tunneling with transmuted material singularity, Phys. Rev. Lett., 125, 20, 207401 (2020).
57. D. C. Adams, S. Inampudi, T. Ribaudo et al., Funneling light through a subwavelength aperture with epsilon-near-zero materials, Phys. Rev. Lett., 107, 13, 133901 (2011).
58. Y. Li and N. Engheta, Supercoupling of surface waves with ε-near-zero metastructures, Phys. Rev. B – Condens. Matter Mater. Phys., 90, 20, 201107 (2014).
59. Z. Li, Y. Sun, H. Sun et al., J. Phys. D. Appl. Phys., 50, 37 (2017).
60. A. Alu and N. Engheta, Coaxial-to-waveguide matching with ε-near-zero ultranarrow channels and bends, IEEE Trans. Antennas Propag., 58, 2, 328–339 (2010).
61. Y. Li, Plasmonic Optics: Theory and Applications (SPIE Press, 2017).
62. P. West, S. Ishii, G. Naik et al., Searching for better plasmonic materials. Las. Photon. Rev. 4, 6, 795–808 (2010).

63. M. Ordal, R. Bell, R. Alexander, Jr, L. Long and M. Querry, Optical properties of fourteen metals in the infrared and far infrared: Al, Co, Cu, Au, Fe, Pb, Mo, Ni, Pd, Pt, Ag, Ti, V, and W., Appl. Opt., 24, 4493 (1985).

64. A. Boltasseva, and H. Atwater, Low-loss plasmonic metamaterials, Science, 331, 290–291 (2011).

65. G. Naik, J. Kim, and A. Boltasseva, Oxides and nitrides as alternative plasmonic materials in the optical range, Opt. Mater. Express, 1, 6, 1090–1099 (2011).

66. S. Vassant, A. Archambault, F. Marquier et al., Epsilon-near-zero mode for active optoelectronic devices. Phys. Rev. Lett., 109, 237401 (2012).

67. S. Campione, I. Brener, and F. Marquier, Theory of epsilon-near-zero modes in ultrathin films. Phys. Rev. B, 91, 12, 121408 (2015).

68. D. de Ceglia, S. Campione, M. A. Vincenti, F. Capolino, M. Scalora, Low-damping epsilon-near-zero slabs: nonlinear and nonlocal optical properties. Phys. Rev. B, 87, 15, 155140 (2013).

69. J. Ou, J.-K. So, G. Adamo et al., Ultraviolet and visible range plasmonics of a topological insulator $Bi_{1.5}Sb_{0.5}Te_{1.8}Se_{1.2}$, Nat. Commun., 5, 5139 (2014).

70. C. Giovampaola and N. Engheta, Plasmonics without negative dielectrics. Phys. Rev. B, 93, 19, 195152 (2016).

71. Y. Li, I. Liberal, and N. Engheta, Structural dispersion–based reduction of loss in epsilon-near-zero and surface plasmon polariton waves, Sci. Adv., 5, eaav3764 (2019).

72. M. Javani and M. Stockman, Real and imaginary properties of epsilon-near-zero materials. Phys. Rev. Lett., 117, 10, 107404 (2016).

73. L. Sun and K. Yu, Strategy for designing broadband epsilon-near-zero metamaterials. J. Opt. Soc. Am., B 29, 5, 984–989 (2012).

74. Z. Li, Z. Liu, and K. Aydin, Wideband zero-index metacrystal with high transmission at visible frequencies, J. Opt. Soc. Am. B., 34, 7, D13–D17 (2017).

75. R. Maas, J. Parsons, N. Engheta, and A. Polman, Experimental realization of an epsilon-near-zero metamaterial at visible wavelengths, Nat. Photon., 7, 11, 907–912 (2013).

76. E. Forati and G. Hanson, On the epsilon near zero condition for spatially dispersive materials, New J. Phys., 15, 12, 123027 (2013).

77. J. Luo, H. Chen, B. Hou, P. Xu, and Y. Lai, Nonlocality-induced negative refraction and subwavelength imaging by parabolic dispersions in metal–dielectric multilayered structures with effective zero permittivity, Plasmonics, 8, 2, 1095–1099 (2013).

78. L. Shen, T. Yang, and Y. Chau, 50/50 beam splitter using a one-dimensional metal photonic crystal with parabolalike dispersion, Appl. Phys. Lett., 90, 25, 251909 (2007).

79. M. Silveirinha, and P. Belov, Spatial dispersion in lattices of split ring resonators with permeability near zero, Phys. Rev., B 77, 23, 233104 (2008).

80. P. Moitra, Y. Yang, Z. Anderson et al., Realization of an all-dielectric zero-index optical metamaterial, Nature Photon., 7, 10, 791–795 (2013).

81. Y. Zhou, X. T. He, F. L. Zhao, and J. W. Dong, Proposal for achieving in-plane magnetic mirrors by silicon photonic crystals, Opt. Lett., 41, 10, 2209–2212 (2016).

82. J. Luo, and Y. Lai, Epsilon-near-zero or mu-near-zero materials composed of dielectric photonic crystals, Sci. China Inf. Sci., 56, 12, 1–10 (2013).

83. H. Iizuka, and N. Engheta, Antireflection structure for an effective refractive index near-zero medium in a two-dimensional photonic crystal, Phys. Rev. B, 90, 11, 115412 (2014).

84. Y. Wu, A semi-Dirac point and an electromagnetic topological transition in a dielectric photonic crystal, Opt Express., 22, 1906–17, 2014.

85. Z. Lin, L. Christakis, Y. Li et al., Topology-optimized dual-polarization Dirac cones, Phys. Rev. B, 97, 08, 081408 (2018).

86. M. Minkov, I. Williamson, M. Xiao, and S. Fan, Zero-index bound states in the continuum, Phys. Rev. Lett., 121, 26, 263901 (2018).

87. H. Tang, C. DeVault, S. Camayd-Muñoz et al., Low-loss zero-index materials, Nano Lett., 21, 2, 914–920 (2021).

88. T. Dong, J. Liang, S. Camayd-Muñoz et al., Ultra-low-loss on-chip zero-index materials, Light-Sci. Appl., 10, 1, 1–9 (2021).

89. M. Selvanayagam and G. Eleftheriades, Negative-refractive-index transmission lines with expanded unit cells, IEEE Trans. Antennas Propag., 56, 11, 3592–3596 (2008).

90. H. Jiang, W. Liu, K. Yu et al., Experimental verification of loss-induced field enhancement and collimation in anisotropic μ-near-zero metamaterials, Phys. Rev. B, 91, 4, 045302 (2015).

91. Y. Li, H. T. Jiang, W. W. Liu et al., Experimental realization of subwavelength flux manipulation in anisotropic near-zero index metamaterials, Europhys. Lett., 113, 5, 57006 (2016).

92. M. Silveirinha, A. Alù, and N. Engheta, Parallel-plate metamaterials for cloaking structures, Phys. Rev. E, 75, 3, 036603-16 (2007).

93. D. M. Pozar, Microwave Engineering, 4th ed. (John Wiley & Sons, Inc., New York, 2012).

94. Y. Cassivi, L. Perregrini, P. Arioni et al., Dispersion characteristics of substrate integrated rectangular waveguide, IEEE Microw. Wireless Compon. Lett., 12, 9, 333–335 (2002).

95. E. Vesseur, T. Coenen, H. Caglayan, N. Engheta, and A. Polman, Experimental verification of n=0 structures for visible light, Physical Review Letters, 110, 1, 013902 (2013).

96. O. Reshef, P. Camayd-Muñoz, D. Vulis et al., Direct observation of phase-free propagation in a silicon waveguide, ACS Photonics, 4, 10, 2385–2389 (2017).

97. J. Luo, W. Lu, Z. Hang et al., Phys. Rev. Lett., 112, 7, 073903 (2014).

98. H. F. Ma, J. H. Shi, Q. Cheng, and T. J. Cui, Experimental verification of supercoupling and cloaking using mu-near-zero materials based on a waveguide, Appl. Phys. Lett., 103, 2, 021908 (2013).

99. J. Luo, P. Xu, H. Chen et al., Realizing almost perfect bending waveguides with anisotropic epsilon-near-zero metamaterials, Appl. Phys. Lett., 100, 22, 221903 (2012).

100. J. Luo, and Y. Lai, Anisotropic zero-index waveguide with arbitrary shapes, Sci. Rep., 4, 1, 5875 (2014).

101. W. Ji, J. Luo, and Y. Lai, Extremely anisotropic epsilon-near-zero media in waveguide metamaterials, Opt. Express, 27, 14, 19463–19473 (2019).

102. A. S. Shalin, P. Ginzburg, A. A. Orlov et al., Scattering suppression from arbitrary objects in spatially dispersive layered metamaterials, Phys. Rev. B, 91, 12, 125426 (2015).

103. J. Luo, P. Xu, L. Gao, Y. Lai, and H. Chen, Manipulate the transmissions using index-near-zero or epsilon-near-zero metamaterials with coated defects, Plasmonics, 7, 2, 353–358 (2012).

104. T. Wang, J. Luo, L. Gao, P. Xu, and Y. Lai, Equivalent perfect magnetic conductor based on epsilon-near-zero media, Appl. Phys. Lett., 104, 21, 211904 (2014).

105. Y. Jin, and S. L. He, Enhancing and suppressing radiation with some permeability-near-zero structures, Opt. Express, 18, 16, 16587–16593 (2010).

106. J. Hao, W. Yan, and M. Qiu, Super-reflection and cloaking based on zero index metamaterial, Appl. Phys. Lett., 96, 10, 101109 (2010).

107. V. C. Nguyen, L. Chen, and K. Halterman, Total transmission and total reflection by zero index metamaterials with defects, Phys. Rev. Lett., 105, 23, 233908 (2010).

108. Y. Xu, and H. Chen, Total reflection and transmission by epsilon-near-zero metamaterials with defects, Appl. Phys. Lett., 98, 11, 113501 (2011).

109. T. Wang, J. Luo, L. Gao, P. Xu, and Y. Lai, Hiding objects and obtaining Fano resonances in index-near-zero and epsilon-near-zero metamaterials with Bragg-fiber-like defects, J. Opt. Soc. Am. B, 30, 7, 1878–1884 (2013).

110. J. Luo, Z. Hang, C. Chan, and Y. Lai, Unusual percolation threshold of electromagnetic waves in double-zero medium embedded with random inclusions, Laser Photonics Rev., 9, 5, 523–529 (2015).

111. J. Luo, B. Liu, Z. Hang, and Y. Lai, Coherent Perfect Absorption via Photonic Doping of Zero-Index Media, Laser Photonics Rev., 12, 8, 1800001 (2018).

112. J. Luo, J. Li, and Y. Lai, Electromagnetic impurity-immunity induced by parity-time symmetry, Phys. Rev. X, 8, 3, 031035 (2018).

113. I. Liberal, Y. Li, and N. Engheta, Reconfigurable epsilon-near-zero metasurfaces via photonic doping, Nanophotonics, 7, 6, 1117–1127 (2018).

114. M. Coppolaro, M. Moccia, G. Castaldi, N. Engheta, and V. Galdi, Non-Hermitian doping of epsilon-near-zero media, Proceedings of the National Academy of Sciences, 117, 25, 13921 (2020).

115. M. Malléjac, A. Merkel, V. Tournat, J.-P. Groby, and V. Romero-García, Doping of a plate-type acoustic metamaterial, Phys. Rev. B, 102, 6, 060302 (2020).

116. A. Sanada, M. Kimura, I. Awai, C. Caloz, and T. Itoh, A planar zeroth-order resonator antenna using a left-handed transmission line, 34th European Microwave Conference, 1341–1344 (2004).

117. J. Kim, G. Kim, W. Seong, and J. Choi, A tunable internal antenna with an epsilon negative zeroth order resonator for DVB-H Service, IEEE Trans. Antennas Propag., 57, 12, 4014–4017 (2009).

118. D. Mitra, B. Ghosh, A. Sarkhel, and S. R. Bhadra Chaudhuri, A miniaturized ring slot antenna design with enhanced radiation characteristics, IEEE Trans. Antennas Propag., 64, 1, 300–305 (2016).

119. S. Jahani, J. Rashed-Mohassel, and M. Shahabadi, Miniaturization of circular patch antennas using MNG metamaterials, , IEEE Antennas Wireless Propag. Lett., 9, 1194–1196 (2010).

120. J. Li, A. Salandrino, and N. Engheta, Shaping light beams in the nanometer scale: a Yagi-Uda nanoantenna in the optical domain. Phys. Rev. B., 76, 24, 245403 (2007).

121. J. Xiong, X. Lin, Y. Yu et al., Novel flexible dual-frequency broadside radiating rectangular patch antennas based on complementary planar ENZ or MNZ metamaterials, IEEE Trans. Antennas Propag., 60, 8, 3958–3961 (2012).

122. J. C. Soric, N. Engheta, S. Maci, and A. Alu, Omnidirectional metamaterial antennas based on ε-near-zero channel matching, IEEE Trans. Antennas Propag., 61, 1, 33–44 (2013).

123. J. Park, Y. Ryu, J. Lee, and J. Lee, Epsilon negative zeroth-order resonator antenna, IEEE Trans. Antennas Propag., 55, 12, 3710–3712 (2007).

124. Z. Zhou and Y. Li, Effective epsilon-near-zero (ENZ) antenna based on transverse cutoff mode, IEEE Trans. Antennas Propag., 67, 4, 2289–2297, (2019).

125. Z. Zhou and Y. Li, A photonic-doping-inspired SIW antenna with length-invariant operating frequency, IEEE Trans. Antennas Propag., 68, 7, 5151–5158 (2020).

126. Z. Hu, C. Chen, Z. Zhou, and Y. Li, An epsilon-near-zero-inspired PDMS substrate antenna with deformation-insensitive operating frequency, IEEE Antennas Wireless Propag. Lett., 19, 9, 1591–1595 (2020).

127. G. Lovat, P. Burghignoli, F. Capolino, D. R. Jackson, and D. R. Wilton, Analysis of directive radiation from a line source in a metamaterial slab with low permittivity, IEEE Trans. Antennas Propag., 54, 3, 1017–1030 (2006).

128. S. Enoch, G. Tayeb, P. Sabouroux, N. Guérin, and P. Vincent, A metamaterial for directive emission, Phys. Rev. Lett., 89, 21, 213902 (2002).

129. E. Forati, G. W. Hanson, and D. F. Sievenpiper, an epsilon-near-zero total-internal-reflection metamaterial antenna, IEEE Trans. Antennas Propag., 63, 5, 1909–1916 (2015).

130. M. Memarian and G. Eleftheriades, Dirac leaky-wave antennas for continuous beam scanning from photonic crystals. Nat. Commun., 6, 5855 (2015).

131. A. H. Dorrah and G. V. Eleftheriades, Pencil-beam single-point-fed Dirac leaky-wave antenna on a transmission-line grid, IEEE Antennas Wireless Propag. Lett., 16, 545–548 (2017).

132. J. Yang, Y. Francescato, S. Maier, F. Mao, and M. Huang, Mu and epsilon near zero metamaterials for perfect coherence and new antenna designs, Opt. Express, 22, 8, 9107–9114 (2014).

133. N. Engheta, C. Papas, and C. Elachi, Radiation patterns of interfacial dipole antennas, Radio Sci., 17, 6, 1557–1566 (1982).

134. D. Mitra, A. Sarkhel, O. Kundu, and S. R. B. Chaudhuri, Design of compact and high directive slot antennas using grounded metamaterial slab, IEEE Antennas Wireless Propag. Lett., 14, 811–814 (2015).

135. A. Dadgarpour, M. Sharifi Sorkherizi, and A. A. Kishk, High-efficient circularly polarized magnetoelectric dipole antenna for 5G applications using dual-polarized split-ring resonator lens, IEEE Trans. Antennas Propag., 65, 8, 4263–4267 (2017).

136. J. Kim et al. Role of epsilon-near-zero substrates in the optical response of plasmonic antennas, Optica, 3, 3, 339–346 (2016).

137. A. Alù, M. G. Silveirinha, A. Salandrino, and N. Engheta, Epsilon-near-zero metamaterials and electromagnetic sources: tailoring the radiation phase pattern, Phys. Rev. B, 75, 15, 155410 (2007).

138. L. Sun, S. Feng, and X. Yang, Loss enhanced transmission and collimation in anisotropic epsilon-near-zero metamaterials, Appl. Phys. Lett., 101, 24, 241101 (2012).

139. S. Feng, Loss-induced omnidirectional bending to the normal in ε-near-zero metamaterials, Phys. Rev. Lett., 108, 19, 193904 (2012).

140. B. Wang and K. Huang, Shaping the radiation pattern with MU and epsilon-near-zero metamaterials, Progress in Electromagnetics Research, 106, 107–119 (2010).

141. M. Navarro-Cía, M. Beruete, I. Campillo, and M. Sorolla, Enhanced lens by ε and μ near-zero metamaterial boosted by extraordinary optical transmission, Phys. Rev. B, 83, 11, 115112 (2011).

142. D. Li, Z. Szabo, X. Qing, E. Li, and Z. N. Chen, A high gain antenna with an optimized metamaterial inspired superstrate, IEEE Trans. Antennas Propag., 60, 12, 6018–6023 (2012).

143. D. Ramaccia, F. Scattone, F. Bilotti, and A. Toscano, Broadband compact horn antennas by using EPS-ENZ metamaterial lens, IEEE Trans. Antennas Propag., 61, 6, 2929–2937 (2013).

144. M. Navarro-Cía, M. Beruete, M. Sorolla, and N. Engheta, Lensing system and Fourier transformation using epsilon-near-zero metamaterials, Phys. Rev. B, 86, 16, 165130 (2012).

145. V. Pacheco-Peña, V. Torres, B. Orazbayev et al., Mechanical 144 GHz beam steering with all-metallic epsilon-near-zero lens antenna, Appl. Phys. Lett., 105, 24, 243503 (2014).

146. J. C. Soric and A. Alù, Longitudinally independent matching and arbitrary wave patterning using ε-near-zero channels, IEEE Trans. Microw. Theory Techn., 63, 11, 3558–3567 (2015).

147. J. C. Soric and A. Alù, Radiation patterning enabled by ε-near-zero reconfigurable metamaterial lenses, 2014 IEEE Antennas and Propagation Society International Symposium (APSURSI), 175–176 (2014).

148. Q. Cheng, W. Jiang, and T. J. Cui , Radiation of planar electromagnetic waves by a line source in anisotropic metamaterials, J. Phys. D: Appl. Phys., 43, 33, 335406 (2010).

149. A. Dadgarpour, M. S. Sorkherizi, T. A. Denidni, and A. A. Kishk, Passive beam switching and dual-beam radiation slot antenna loaded with ENZ medium and excited through ridge gap waveguide at millimeter-waves, IEEE Trans. Antennas Propag., 65, 1, 92–102 (2017).

150. G. Castaldi, S. Savoia, V. Galdi, A. Alù, and N. Engheta, Analytical study of subwavelength imaging by uniaxial epsilon-near-zero metamaterial slabs, Phys. Rev. B, 86, 11, 115123 (2012).

151. V. Pacheco-Peña, N. Engheta, S. Kuznetsov, A. Gentselev, and M. Beruete, Experimental realization of an epsilon-near-zero graded-index metalens at terahertz frequencies, Phys. Rev. Applied, 8, 3, 034036 (2017).

152. Y. Jin, S. Xiao, N. Asger Mortensen, and S. He, Arbitrarily thin metamaterial structure for perfect absorption and giant magnification, Opt. Express, 19, 12, 11114–11119 (2011).

153. S. Zhong, Y. Ma, and S. He, Perfect absorption in ultrathin anisotropic ε-near-zero metamaterials, Appl. Phys. Lett., 105, 2, 023504 (2014).

154. J. Park, J. H. Kang, X. Liu, and M. L. Brongersma, Electrically tunable Epsilon-Near-Zero (ENZ) metafilm absorbers, Sci Rep., 5, 15754 (2015).

155. Y. Kato, S. Morita, H. Shiomi, and A. Sanada, Ultrathin perfect absorbers for normal incident waves using Dirac cone metasurfaces with critical external coupling, IEEE Microw. Wireless Compon. Lett., 30, 4, 383–386 (2020).

156. A. Anopchenko, L. Tao, C. Arndt, and H. W. H. Lee, Field-effect tunable and broadband epsilon-near-zero perfect absorbers with deep subwavelength thickness, ACS Photonic., 5, 7, 2631 (2018).

157. D. A. Powell, A. Alù, B. Edwards et al., Nonlinear control of tunneling through an epsilon-near-zero channel, Phys. Rev. B, 79, 24, 245135 (2009).

158. N. Vojnović, B. Jokanović, M. Mitrović, F. Mesa, and F. Medina, Tuning ZOR in ENZ waveguide using a single longitudinal slot and equivalent circuit parameter extraction, 2014 8th International Congress on Advanced Electromagnetic Materials in Microwaves and Optics, 283–285 (2014).

159. Y. He, Y. Li, L. Zhu et al., Waveguide dispersion tailoring by using embedded impedance surfaces, Phys. Rev. Applied, 10, 6, 064024 (2018).

160. Y. He, Y. Li, Z. Zhou et al., Wideband epsilon-near-zero supercoupling control through substrate-integrated impedance surface, Adv. Theory Simul., 2, 8, 1900059 (2019).

161. string-name>A. Alù and N. Engheta, Dielectric sensing in ε-near-zero narrow waveguide channels, Phys. Rev. B, 78, 4, 045102 (2008).

162. H. Lobato-Morales, D. V. B. Murthy, A. Corona-Chavez et al., Permittivity measurements at microwave frequencies using Epsilon-Near-Zero (ENZ) tunnel structure, IEEE Trans. Microw. Theory Techn., 59, 7, 1863–1868 (2011).

163. A. Corona-Chavez, D. V. B. Murthy, and J. L. Olvera-Cervantes, Novel microwave filters based on epsilon near zero waveguide tunnels, Microw. Optical Technol. Lett., 53, 8, 1706–1710 (2011).

164. M. Radovanovic and B. Jokanovic, Dual-band filter inspired by ENZ waveguide, IEEE Microw. Wireless Compon. Lett., 27, 6, 554–556 (2017).

165. Z. Zhou, Y. Li, E. Nahvi et al., Phys. Rev. Applied, 13, 3, 034005 (2020).

166. Y. Wang, P. Xu, and Y. Qin, Optical-phase demodulation using zero-index metamaterials, Opt. Lett., 40, 13, 3157–3160 (2015).

167. I. Liberal, and N. Engheta, Multiqubit subradiant states in N-port waveguide devices: ε-and-μ-near-zero hubs and nonreciprocal circulators, Phys. Rev. A, 97, 2, 022309 (2018).

168. Z. Zhou and Y. Li, N-port equal/unequal-split power dividers using epsilon-near-zero metamaterials, IEEE Trans. Microw. Theory Techn., 69, 3, 1529–1537 (2021).

169. M. Silveirinha, Trapping light in open plasmonic nanostructures, Phys. Rev. A, 89, 2, 023813(2014).

170. I. Liberal, and N. Engheta, Nonradiating and radiating modes excited by quantum emitters in open epsilon-near-zero cavities, Sci. Adv., 2, 10, e1600987 (2016).

171. L. Li, J. Zhang, C. Wang, N. Zheng, and H. Yin, Optical bound states in the continuum in a single slab with zero refractive index, Phys. Rev. A, 96, 1, 013801 (2017).

172. S. Lannebère, and M. Silveirinha, Optical meta-atom for localization of light with quantized energy, Nat Commun., 6, 8766 (2015).

173. J. Huang, X. Zhang, L. Zhang, and J. Zhang, General model of optical frequency conversion in homogeneous media: application to second-harmonic generation in an ε-near-zero waveguide, Phys. Rev. A, 96, 1, 013836 (2017).

174. A. Ciattoni, C. Rizza, and E. Palange, Extreme nonlinear electrodynamics in metamaterials with very small linear dielectric permittivity, Phys. Rev. A, 81, 4, 043839 (2010).

175. C. Argyropoulos, P. Y. Chen, G. D'Aguanno, N. Engheta, and A. Alù, Boosting optical nonlinearities in ε-near-zero plasmonic channels, Phys. Rev. B, 85, 4, 045129 (2012).

176. Y. Yang, J. Lu, A. Manjavacas, et al. High-harmonic generation from an epsilon-near-zero material, Nat. Phys., 15, 1022–1026 (2019).

177. A. Capretti, Y. Wang, N. Engheta, and L. Negro, Enhanced third-harmonic generation in Si-compatible epsilon-near-zero indium tin oxide nanolayers, Opt. Lett., 40, 7, 1500–1503 (2015).

178. M. A. Vincenti, M. Kamandi, D. de Ceglia et al., Second-harmonic generation in longitudinal epsilon-near-zero materials, Phys. Rev. B, 96, 4, 045438 (2017).

179. M. Z. Alam, I. De Leon, and R. W. Boyd, Large optical nonlinearity of indium tin oxide in its epsilon-near-zero region, Science, 352, 6287, 795–797 (2016).

180. N. Kinsey, C. DeVault, J. Kim et al., Epsilon-near-zero Al-doped ZnO for ultrafast switching at telecom wavelengths, Optica, 2, 7, 616–622 (2015).

181. Q. Guo, Y. Cui, Y. Yao et al., A solution-processed ultrafast optical switch based on a nanostructured epsilon-near-zero medium, Adv. Mater., 29, 27, 1700754 (2017).

182. T. Tanabe, M. Notomi, S. Mitsugi, A. Shinya, and E. Kuramochi, Fast bistable all-optical switch and memory on a silicon photonic crystal on-chip, Opt. Lett., 30, 19, 2575–2577 (2005).

183. P. A. Belov, R. Marqués, S. I. Maslovski et al., Strong spatial dispersion in wire media in the very large wavelength limit, Phys. Rev. B, 67, 11, 113103 (2003).

184. N. Engheta, From RF circuits to optical nanocircuits, IEEE Microw. Mag., 13, 4, 100–113, May–June (2012).

185. N. Engheta, Taming light at the nanoscale, Phys. World, 13, 9, 31–34 (2010).

186. B. Edwards and N. Engheta, Experimental verification of displacement-current conduits in metamaterials-inspired optical circuitry, Phys. Rev. Lett., 108, 19, 193902 (2012).

187. A. Alù, M. E. Young, and N. Engheta, Design of nanofilters for optical nanocircuits, Phys. Rev. B, 77, 14, 144107 (2008).

188. Y. Li, I. Liberal, and N. Engheta, Dispersion synthesis with multi-ordered metatronic filters, Opt. Express, 25, 3, 1937–1948 (2017).

189. Y. Li and Z. Zhang, Experimental verification of guided-wave lumped circuits using waveguide metamaterials, Phy. Rev. Appl., 9, 4, 044024 (2018).

190. O. Dominguez, L. Nordin, J. Lu et al., Monochromatic multimode antennas on epsilon-near-zero materials, Adv. Opt. Mater., 7, 10, 1800826 (2019).

191. Z. Chai, X. Hu, F. Wang et al., Ultrafast on-chip remotely-triggered all-optical switching based on epsilon-near-zero nanocomposites, Laser Photonics Rev., 11, 5, 1700042 (2017).

192. Y. Wu, X. Hu, F. Wang et al., Ultracompact and unidirectional on-chip light source based on epsilon-near-zero materials in an optical communication range, Phys. Rev. Applied, 12, 5, 054021 (2019).

193. S. Feng, and K. Halterman, Coherent perfect absorption in epsilon-near-zero metamaterials, Phys. Rev. B, 86, 16, 165103 (2012).

194. J. Luo, S. Li, B. Hou, and Y. Lai, Unified theory for perfect absorption in ultrathin absorptive films with constant tangential electric or magnetic fields, Phys. Rev. B, 90, 16, 165128 (2014).

195. D. Wang, J. Luo, Z. Sun, and Y. Lai, Transforming zero-index media into geometry-invariant coherent perfect absorbers via embedded conductive films, Opt. Express, 29, 4, 5247 (2021).

196. S. Feng, Loss-induced omnidirectional bending to the normal in ε-near-zero metamaterials, Phys. Rev. Lett., 108, 19, 193904 (2012).

197. V. Y. Fedorov, and T. Nakajima, All-angle collimation of incident light in mu-near-zero metamaterials, Opt. Express, 21, 23, 27789–27795 (2013).

198. A. Alù and N. Engheta, Boosting molecular fluorescence with a plasmonic nanolauncher. Phys. Rev. Lett., 103, 4, 043902 (2009).

199. R. Sokhoyan and H. A. Atwater, Quantum optical properties of a dipole emitter coupled to an ε-near-zero nanoscale waveguide, Opt. Express, 21, 26, 32279–32290 (2013).

200. I. Liberal and N. Engheta, Zero-index structures as an alternative platform for quantum optics, PNAS, 114, 5, 822–827 (2017).

201. S.-A. Biehs and G. S. Agarwal, Qubit entanglement across ε-near-zero media, Phys. Rev. A, 96, 2, 022308 (2017)

202. L. Vertchenko, N. Akopian, and A. V. Lavrinenko, Epsilon-near-zero grids for on-chip quantum networks. Sci. Rep., 9, 6053 (2019)

203. F. J. Rodriguez-Fortuo, A. Vakil, and N. Engheta, Electric levitation using ε -near-zero metamaterials. Phys. Rev. Lett., 112, 3, 033902 (2014).

204. I. Liberal and N. Engheta, Manipulating thermal emission with spatially static fluctuating fields in arbitrarily shaped epsilon-near-zero bodies, PNAS, 115, 12, 2878–2883 (2018).

205. I. Liberal, M. Lobet, Y. Li, and N. Engheta, Near-zero-index media as electromagnetic ideal fluids, PNAS, 117, 39, 24050–24054 (2020).

206. H. C. Chu, Q. Li, B. Liu et al., A hybrid invisibility cloak based on integration of transparent metasurfaces and zero-index materials, Light: Science & Applications, 7, 50 (2018).

207. L. Sun, J. Gao, and X. Yang, Broadband epsilon-near-zero metamaterials with steplike metal–dielectric multilayer structures, Phys. Rev. B, 87, 16, 165134 (2013).

208. L. Sun, K. W. Yu, and G. P. Wang, Design anisotropic broadband ε-near-zero metamaterials: rigorous use of Bergman and Milton spectral representations, Phys. Rev. Applied, 9, 6, 064020 (2018).

209. V. Lucarini, J. J. Saarinen, K.-E. Peiponen, and E. M. Vartiainen, Kramers–Kronig relations in optical materials research (Springer, 2005).

210. F. Zangeneh-Nejad, D. L. Sounas, A. Alù et al. Analogue computing with metamaterials. Nat. Rev. Mater., 6, 207–225 (2021).

Cambridge Elements ≡

Emerging Theories and Technologies in Metamaterials

Tie Jun Cui

Southeast University, China

Tie Jun Cui is Cheung-Kong Professor and Chief Professor at Southeast University, China, and a Fellow of the IEEE. He has made significant contributions to the area of effective-medium metamaterials and spoof surface plasmon polaritons at microwave frequencies, both in new-physics verification and engineering applications. He has recently proposed digital coding, field-programmable, and information metamaterials, which extend the concept of metamaterial.

John B. Pendry

Imperial College London

Sir John Pendry is Chair in Theoretical Solid State Physics at Imperial College London, and a Fellow of the Royal Society, the Institute of Physics and the Optical Society of America. Among his many achievements are the proposal of the concepts of an 'invisibility cloak' and the invention of the transformation optics technique for the control of electromagnetic fields.

About the Series

This series systematically covers the theory, characterisation, design and fabrication of metamaterials in areas such as electromagnetics and optics, plasmonics, acoustics and thermal science, nanoelectronics, and nanophotonics, and also showcases the very latest experimental techniques and applications. Presenting cutting-edge research and novel results in a timely, indepth and yet digestible way, it is perfect for graduate students, researchers, and professionals working on metamaterials.

Cambridge Elements ≡

Emerging Theories and Technologies in Metamaterials

Printed in the United States
by Baker & Taylor Publisher Services